D0758842

Scalable Video on Demand

Scalable Video on Demand

Adaptive Internet-based Distribution

Michael Zink
University of Massachusetts, Amherst, USA

Telephone (+44) 1243 779777

Email (for orders and customer service enquiries): cs-books@wiley.co.uk
Visit our Home Page on www.wiley.com

Other Wiley Editorial Offices

John Wiley & Sons Inc., 111 River Street, Hoboken, NJ 07030, USA

Jossey-Bass, 989 Market Street, San Francisco, CA 94103-1741, USA

Wiley–VCH Verlag GmbH, Boschstr. 12, D-69469 Weinheim, Germany

John Wiley & Sons Australia Ltd, 42 McDougall Street, Milton, Queensland 4064, Australia

John Wiley & Sons (Asia) Pte Ltd, 2 Clementi Loop #02-01, Jin Xing Distripark, Singapore 129809

John Wiley & Sons Canada Ltd, 22 Worcester Road, Etobicoke, Ontario, Canada M9W 1L1

Wiley also publishes its books in a variety of electronic formats. Some content that appears in print may not be available in electronic books.

British Library Cataloguing in Publication Data

A catalogue record for this book is available from the British Library

ISBN-13 978-0-470-02268-9 (HB)
ISBN-10 0-470-02268-X (HB)

Typeset in 10.5/13pt Times by Integra Software Services Pvt. Ltd, Pondicherry, India
Printed and bound in Great Britain by TJ International Padstow, Cornwall
This book is printed on acid-free paper responsibly manufactured from sustainable forestry in which at least two trees are planted for each one used for paper production.

To Andrea

Contents

List of Figures

List of Tables

About the Author

Michael Zink is currently a postdoctoral fellow in the Computer Science Department at the University of Massachusetts in Amherst. Before, he worked as a researcher at the Multimedia Communications Lab at Darmstadt University of Technology. He works in the fields of sensor networks and distribution networks for high bandwidth data. Further research interests are in wide-area multimedia distribution for wired and wireless environments and network protocols. He is one of the developers of the KOMSSYS streaming platform. He received his Diploma (MSc) from Darmstadt University of Technology in 1997. From 1997 to 1998 he was employed as guest researcher at the National Institute of Standards and Technology (NIST) in Gaithersburg, MD, where he developed an MPLS testbed. In 2003, he received his PhD degree (Dr.-Ing.) from Darmstadt University of Technology, his thesis was on "Scalable Internet Video-on-Demand Systems".

Acknowledgements

First of all I would like to thank my parents. Without their support and love this work would never have been possible. Also to my sister Christina, and her family Oli, Niklas, and Marla-Marie for always being there for me and reminding me how beautiful the simple things in life can be.

This book is based on work I did for my dissertation at the Multimedia Communications Lab at Darmstadt University of Technology.

A sincere thank you goes to my advisor Ralf Steinmetz for providing the environment in which I could carry out my research and for giving me the opportunity to learn many more things than I originally expected.

I also owe a great deal to my friend Carsten Griwodz. The fruitful discussions I could lead with him initiated many of the results presented in this book. I am also thankful for the long lasting colaboration between the two of us. His contributions to Chapters 8 and 9 and all his work for KOMSSYS are invaluable.

Jens Schmitt never gave up in motivating me by showing me the meaning of scientific work. Thank you.

I also want to thank my colleagues Utz, Lipi, Nicole, Ralf, Kalli, Andreas, Krishna, Andreas, and Giwon, who accompanied me on the journey of my dissertation in Darmstadt.

David Cypher has done an incredible job in helping me to improve my technical writing and Moni Jayme allowed me to use some of the resources in Darmstadt after I left for my new destination.

I finished the manuscript for this book after I started working in the Computer Science Department of the University of Massachusetts in Amherst. I enjoy working in this place a lot! Next to all my great colleagues, I especially want to thank Jim Kurose and Mark Preston for their support and guidance.

I also want to thank the team at Wiley who did an outstanding job in supporting me during the creation of this book: Emily Bone, Wendy Hunter, Mike Shardlow, Kathryn Sharples and Birgit Gruber.

"The journey is the reward" – Taoist Saying

Acronyms

ADSL	Asymmetric digital subscriber line
AIMD	Additive increase/multiplicative decrease
A/V	Audio/video
AVT	Audio/video transport
BSD	Berkeley Software Distribution
CBR	Constant bit rate
CC	Cache centric
CDI	Content distribution internetworking
CDN	Content distribution network
CFVC	Cache-friendly viewer centric
CPU	Central processing unit
DAVIC	Digital Audio Visual Council
DCCP	Datagram congestion control protocol
DCT	Discrete cosine transformation
DRM	Digital rights management
DSBV	Double stimulus binary vote
DSCQE	Double stimulus continuous quality evaluation
DSIS	Double stimulus impairment scale
DSL	Digital subscriber line
DSM-CC	Digital storage media command and control
DSS	Dynamic stream switching
DVB	Digital video broadcast
ERA	Expanding ring advertisement
EZW	Embedded zerotree wavelet
FEC	Forward error correction
FGS	Fine granularity scalability
FSC	Fair share claiming
FTP	File Transfer Protocol
GM	Graph manager
HTML	Hypertext Markup Language

HTTP	Hypertext Transfer Protocol
ICP	Internet Cache Protocol
IETF	Internet Engineering Task Force
IP	Internet Protocol
ITU	International Telecommunications Union
KDE	K desktop environment
LAN	Local area network
LC-RTP	Loss collection RTP
LD	Layer dummy
LRMP	Light-weight Reliable Multicast Protocol
MAN	Metropolitan area network
MCL	Maximum of cached layers
MDC	Multiple description coding
MMUSIC	Multiparty multimedia session control
MOL	Maximum of original layers
MP3	MPEG-1 Audio Layer III
MPEG	Motion Picture Expert Group
MTU	Maximum transferable unit
NoVoD	No video on demand
NVoD	Near video on demand
OS	Operating system
PDA	Personal digital assistant
POSIX	Portable operating system interface
PPV	Pay-per-view
PSNR	Peak signal-to-noise ratio
PSTN	Public switched telephony network
QoS	Quality of service
QVoD	Quasi video on demand
RAP	Rate adaptation protocol
RAM	Random access memory
RDTSC	Read time stamp counter
RFC	Request for comment
RLM	Receiver-driven layered multicast
RMTP	Reliable Multicast Transport Protocol
RSVP	Resource Reservation Protocol
RTP	Real-time Transport Protocol
RTCP	Real-time Transport Control Protocol
RTSP	Real-time Streaming Protocol
RTT	Round trip time
SAS	Scalable adaptive streaming

SC	Stimulus comparison
SCTP	Stream Control Transmission Protocol
SDP	Session Description Protocol
SH	Stream handler
SNR	Signal-to-noise ratio
SPEG	Scalable MPEG
SRM	Scalable Reliable Multicast
SR-RTP	Scalable Reliable RTP
SSCQE	Single stimulus quality evaluation
TCP	Transmission Control Protocol
TEAR	TCP emulation at receivers
TFMCC	TCP-friendly multicast congestion control
TFRC	TCP-friendly rate control
TRM	Transport Protocol for Reliable Multicast
TTL	Time-to-live
TVoD	True video on demand
UDP	User Datagram Protocol
U-LLF	Unrestricted lowest layer first
U-LL-SGF	Unrestricted lowest layer shortest gap first
UMTS	Universal mobile telecommunication system
U-SG-LLF	Unrestricted shortest gap lowest layer first
VBR	Variable bitrate
VC	Viewer centric
VoD	Video on demand
WLAN	Wireless LAN
W-LLF	Window-based lowest layer first
WWW	World Wide Web
XDSL	Protocols under the DSL umbrella (e.g., ASDL)

1

Introduction

In answer to the increasing popularity of the Internet and especially one of its major applications, the World Wide Web, not only in research and professional environments but also among the wider public, the provision of new types of content is increasing rapidly. In the beginning of the World Wide Web, the available content was mainly based on text (hyperlink) documents and still images. With the increasing popularity for private users, content which is usually provided by the entertainment industry becomes more and more interesting. Such content is usually characterized as multimedia data [1] such as audio, video and the combination of both. With the need for such types of content, new services and applications are offered on the Internet. An example of new services is radio station programmes which are offered, in addition to the traditional terrestrial broadcast, via the Internet.

Nevertheless, it is obvious that these services are only popular where no alternative exists to obtain this content or where the offered quality is comparable to existing alternatives. For example, listening to a radio station via the Internet can be compared to listening to the same station on a car radio. The quality is definitely not equivalent to CD quality: on both Internet and car radio, interruptions and changes in the quality can occur. A client who is used to listening to the radio in his or her car accepts these quality degradations and, thus, is willing to use a service such as Internet radio. The explicit provision of live events on the Internet is an example of where there is a lack of alternatives. In this case users accept a degradation in quality, since there is only the one possibility for receiving the live event. The broadcast of live events, such as pop concerts or sports events, can even lead to partial collapse of the service caused by the high user demand.

The situation is completely different in cases in which alternatives in a much better quality are available. Since the mid-1990s several video on

Scalable Video on Demand: Adaptive Internet-based Distribution M. Zink

demand (VoD) trials have been performed but none of them has resulted in a major success. In fact, there are only a few VoD services available in the Internet. Despite the failure of these trials a huge effort has been put into overcoming the problems that prevent VoD and video streaming from becoming a successful service in the Internet. Recent developments show that VoD is, at least in some areas, gaining popularity. This tendency is certainly supported by new technologies that allow users at home to receive data at a higher bandwidth and, thus, better quality. Yet there are still open issues that have not been solved so far. Two of these issues, which are considered in this book, are the absence of quality of service (QoS) in the Internet and the heterogeneity of the clients. Both require new mechanisms that allow an adaptation of the streaming rate to available network and client resources. Furthermore, an integration of these new mechanisms with a video distribution architecture is necessary to increase the overall scalability of VoD services. In this book, these new mechanisms and their integration in a video distribution architecture are presented.

1.1 WHY SCALABLE INTERNET VIDEO ON DEMAND SYSTEMS?

In the last few years, the Internet has been used for an increasing amount of traffic stemming from the emergence of multimedia applications which use audio and video streaming [2]. This increase is expected to continue and be reinforced since access technologies such as Asymmetric Digital Subscriber Line (ADSL) and cable modems enable residential users to receive high-bandwidth multimedia streams. One specific application which will be enabled by future access technologies is video on demand (VoD). True VoD (TVoD) [3] is a subtype of VoD which allows users to watch a certain video at any desired point in time while also offering the same functions as a standard VCR (i.e., fast-forward, rewind, pause, stop). The challenges of providing TVoD in the Internet are manifold and require the orchestration of different technologies. Some of these technologies, such as video encoding (for example, MPEG-1), are fairly well understood and established. Other technologies such as the distribution and caching of video content and the adaptation of streaming mechanisms to the current network situation and user preferences are still under investigation.

Existing work on TVoD has shown caches to be extremely important with respect to *scalability*, from the network, as well as from the video servers' perspective [4]. Scalability, of course, is an important issue if a TVoD system is to be considered for use in the Internet. Yet, simply reusing concepts from traditional Internet Web caching is not sufficient to suit the special needs of video content since, for example, popularity life cycles can be very different [5].

In addition to scalability, it is very important for an Internet TVoD system to take the 'social' rules implied by Transmission Control Protocol's (TCP) cooperative resource management model into account, i.e., to be *adaptive* in the face of (incipient) network congestion. Therefore, the streaming mechanisms of an Internet TVoD system need to incorporate end-to-end congestion control mechanisms to prevent unfairness against TCP-based traffic and to increase the overall utilization of the network. Note that traditional video streaming mechanisms rely on open-loop control mechanisms, i.e., on explicit reservation and allocation of resources. As it is debatable whether such mechanisms will ever be used in the global Internet, e.g., in the form of RSVP/IntServ [6], mechanisms presented in this book build upon the current best-effort service model of the Internet which is based on closed-loop control exerted by TCP-like congestion control. Yet, since video transmissions need to be paced at their 'natural' rate, adaptiveness can only be integrated into streaming mechanisms in the form of quality degradation and not by delaying the transfer as is possible with elastic traffic such as File Transfer Protocol (FTP) transfers. An elegant way of introducing adaptiveness into streaming is to use scalable video [7] formats as they allow dropping segments (the transfer units) of the video in a controlled way without high computational effort of, for example, adaptive encoding as described in reference [8]. Thus, it overcomes the inelastic characteristics of traditional encoding formats such as MPEG-1 or H.261. In addition, adaptive streaming in combination with an adaptive encoding format like layer-encoded video can avoid uncontrolled losses and, thus, increase the perceived quality of a video in contrast to an uncontrolled streaming. A side-effect of adaptive streaming is the fact that heterogeneous clients and access networks can be supported more efficiently.

Little work has been performed so far on the aspect of combining both, scalability for VoD systems and adaptive streaming. Thus, the focus of this book is on new mechanisms that combine the benefits of both approaches in order to maximize the quality of the video stream that is delivered to the client. However, while the combination of caching and adaptive streaming promises a scalable and TCP-friendly TVoD system, it also creates new design challenges. One drawback of adaptive transmissions is the introduction of quality variations during a streaming session. These variations affect both the viewer's perceived quality and the quality of the cached video and, thus, the acceptance of a service that is based on such technology.

The overall question this book tries to answer is: Can the benefits of system scalability and adaptive streaming be combined to create new systems that can increase the performance of VoD services?

1.2 WHAT IS THE GOAL OF THIS BOOK?

The goal of this book is to answer the aforementioned question by extending existing mechanisms and creating new ones to increase the performance of VoD systems in today's Internet (an Internet without quality of service support). The validity and applicability of these mechanisms are proven through investigations based on assessment, simulation and a prototype implementation which in combination lead to the final results. These new mechanisms should be usable as building blocks for scalable Internet VoD systems. Next to the development of the individual mechanisms it should also be shown how these mechanisms can be orchestrated to build a well-suited distribution infrastructure for VoD services which is in contrast to approaches where only isolated parts of the distribution infrastructure are investigated. Nevertheless, the mechanisms should also be usable independently of each other to allow VoD operators to tailor a service based on these mechanisms according to their specific needs. For example, the mechanism that reduces quality variations, which is located on the cache, should be independent of the transport mechanisms between server and cache. It is certainly not the goal of this book to demonstrate how a specific VoD application is built.

It is a fact that in an Internet without QoS support quality variations and data loss during a streaming session cannot be avoided. Therefore, those quality variations should be kept to a minimum in order to increase the acceptance of VoD services. The minimum of quality variations can also be seen as the maximum number of variations that the viewers tolerate. As a consequence, an intolerable number of variations would lead to the fact that users do not accept the offered service. Investigations on the subjective impressions of quality variations in layer-encoded videos which might reveal such information have not till now been performed. Therefore, this book offers better insight into how variations in scalable video affect the viewers' perceived quality by conducting such a subjective assessment.

Based on this newly gained knowledge we investigate whether caches can be used to improve the quality of a layer-encoded video stream that is transported from or through the cache to the client. In other words, how can these layer variations, with the aid of caches, be kept to a minimum. As a constraint, the mechanisms at the client used to receive and display the video should be kept unchanged. This decision is based on the fact that it is far easier to establish new mechanisms on a manageable number of servers and caches (compared to the enormous number of uncontrollable clients). Next, the new mechanisms should be designed in a way that allows for heterogeneous clients and access networks.

Since scalability in a VoD service means, among other things, to increase the number of simultaneously served clients, it is important that server load is reduced by streaming data directly from caches to the client. Thus, it is also important not only to minimize layer variations in the stream delivered to the client but also to minimize these variations in the cached version of the video to allow the delivery of a high-quality video from the cache. Figure 1.1 shows the elements of a VoD service and the systems that are focused on in this book.

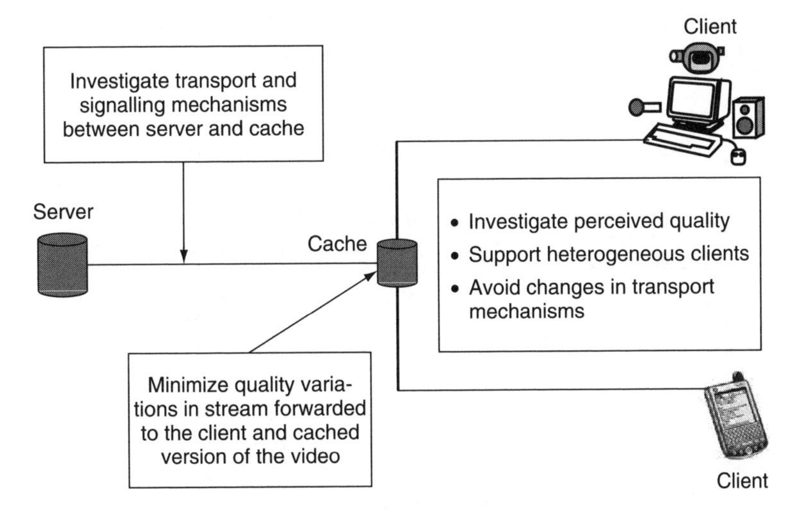

Figure 1.1 Conceptual overview.

1.3 OUTLINE OF THIS BOOK

Chapter 2 gives an overview of the scalable adaptive streaming (SAS) architecture which combines system and content scalability in order to allow VoD services in a best-effort Internet in combination with heterogeneous access networks and clients. After an overview of the SAS architecture the building blocks of such an architecture, which repeatedly occur throughout the book, are introduced.

In Chapter 3 an overview of related work in the area of video streaming and distribution is given. It is shown which solutions are already available as commercial products and what their shortcomings are. Activities in several standardization organizations are briefly mentioned in order to give the interested reader better orientation, since it is not just one organization that is involved. In the remainder of this chapter related work from the research area is presented that has served as the basis for the work presented in this book.

A survey on related work in the area of retransmission scheduling revealed a lack of subjective investigations on how layer variations in layer-encoded video influence the viewer's perceived quality. Existing work on retransmission scheduling is based on speculative assumptions. Therefore, as a first consequence a subjective assessment is performed in the scope of this book to get better insight into how layer variations affect the perceived quality. The subjective assessment is presented in Chapter 4.

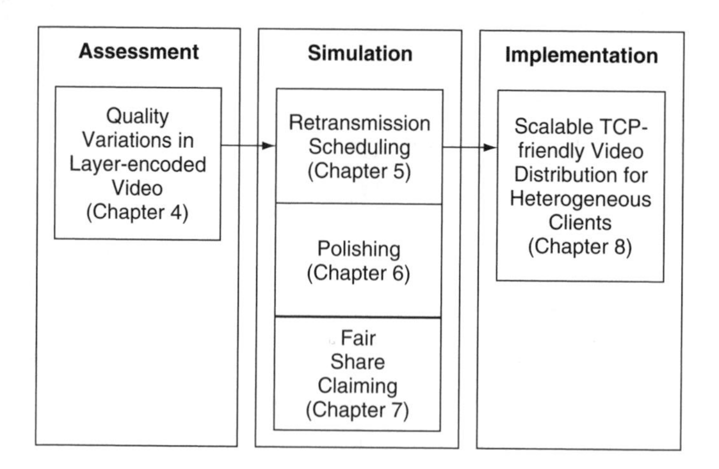

Figure 1.2 Organization of this book.

The results of this investigation are used to confirm the applicability of an objective metric which is developed in order to evaluate heuristics for retransmission scheduling. The investigations on the latter (presented in Chapter 5) show that an optimal solution, given reasonable computing power, is computationally infeasible.

Additionally, the results of the subjective assessment reveal that an increase in the amount of stored data for a cached layer-encoded video object does not necessarily increase its perceived quality (see section 4.5). This means that dropping certain segments of a layer to reduce the amount of variation can increase the perceived quality. Based on this knowledge, reducing layer variations by dropping certain segments seems to be an additional option to improve the perceived quality, leading to a new mechanism called *polishing* which is presented in Chapter 6. Polishing can be used either for cache replacement or during playout from the cache to the client. In the first case, segments are deleted from the cache, based on the polishing algorithm, in order to free storage space for new video objects, while in the second case certain segments are not streamed from the cache to the client.

With a simulative environment solely built to investigate the newly created mechanisms for retransmission scheduling and polishing, a series of simulations are performed. The goal of the simulations is to show the applicability of both mechanisms and their dependence on certain parameters (e.g., the available bandwidth for retransmissions). The results obtained by the simulations were satisfying and showed, in the case of retransmission scheduling, a significant improvement compared to already existing mechanisms.

In subsequent work, which is presented in Chapter 7, a new mechanism allowing the transport of segments requested for retransmission is developed, leading to a combination of TCP-friendly streaming and retransmission scheduling. This approach has a side-effect allowing a TCP-friendly transport stream to claim its fair share on the network path, although layer-encoded video is transmitted. Based on the mechanism for fair share claiming an implementation design for an already existing streaming platform is made.

In Chapter 8, this design is extended to allow scalable TCP-friendly video distribution for heterogeneous clients. Therefore, the cache is extended by gateway functionality enabling standard clients in the SAS architecture. Based on this extended platform, experiments are performed to demonstrate the applicability of the newly created mechanisms which are building blocks of the SAS architecture.

Finally, a summary of the contributions created in this book is given and final conclusions are drawn.

1.4 WHO IS THIS BOOK FOR?

- Students: This book is certainly not created as classic textbook. So, it is not meant for students who are about to learn the basics of video streaming and video distribution systems in the Inter-net. It should rather be seen as introductory literature for students working on streaming-related projects.

In addition, it can serve as complementary literature for students who are highly interested in this topic.

- Lecturers: As already mentioned above, this book does not have textbook character and was not written with this goal in mind. Nevertheless, this book can serve lecturers as additional material to update or extend their existing lectures. For example, parts of this book fit very well in a content distribution lecture, while other parts can be useful in a networking class.
- Designers: This book should be very helpful for software engineers who are involved in the design of a video streaming architecture which is based on existing Internet technology. Many of the results presented in this book can be used as guidelines for making decisions for a future system.
- Implementors: The overall architecture for scalable and adaptive streaming systems presented in this book might be a good starting point for someone who has taken on the task of implementing such a system. It should be mentioned here that most of the results presented in this book are based on an existing implementation which is available as open source (http://komssys.source-forge.net).

2

Scalable Adaptive Streaming Architecture

This chapter gives a general overview of the scalable adaptive streaming (SAS) architecture. SAS allows VoD services in today's Internet which is characterized by best-effort service and a wide heterogeneity regarding end-systems and access networks.

In section 2.1 it is shown that a video distribution system is a special subclass of distributed systems. In a more general sense scalability issues of distributed systems are discussed and mapped to the specific scenario of video distribution. Since replication and its subclasses are means to increase scalability in distributed systems, they are introduced and discussed within the scope of video distribution systems in section 2.2. Terminology for elements used in a video distribution system is not consistent in the literature. To avoid confusion, the terminology used throughout this book is introduced in section 2.3. Section 2.4 gives a detailed presentation of the architecture that is based on the new mechanisms presented in this book. It is shown how video distribution without these mechanisms has been performed so far. Additionally, the drawbacks of such an approach are emphasized. Furthermore, the advantages of caching based on the access characteristics for videos which lead to load reduction on the origin server and an increased system scalability are presented. Subsequently the two major building blocks of the SAS architecture, system and content scalability, are introduced, and the benefits of combining both, system and content scalability, are presented. A typical video distribution scenario, reflecting the situation in the current Internet, is given in section 2.5 to clarify why it is important to combine system and content scalability in the SAS architecture. To emphasize the positive effects of combining both, system and content scalability, an example application for SAS is presented in section 2.6.

Scalable Video on Demand: Adaptive Internet-based Distribution M. Zink

2.1 DISTRIBUTED SYSTEMS

In this section, it is proposed that a video distribution system can be seen as a subclass of distributed systems as they are known in traditional computer science. A distributed system can be defined as a '*collection of independent computers that appears to its users as a single coherent system*' [9]. The applicability of this definition to video distribution systems can be demonstrated by the following example. If we assume a user wants to watch a specific video, the user might access that video through a link on a web page. By choosing that link the video client application is started, which, transparently to the user, decides where to retrieve the video data from. Thus, the user is completely unaware of whether the video is located on a video server, a cache, or even its local disk.

Applications based on distributed systems are manifold and a video distribution system is only one possible example. Other prominent examples are the World Wide Web (WWW), distributed file systems such as CODA [10] or applications used for GRID computing [11]. In a more general way one can speak of a distributed system if several computing devices are cooperating with each other.

One major issue of distributed systems is scalability. A distributed system that scales can deal with an increasing number of users while its growth keeps performance degradation and administrative complexity to a minimum.

In the case of video distribution systems there are three major classes of scalability as shown in the following:

- *User scalability*: A user scalable distributed system allows one to add more users and resources to the system. A good example for *user scalability* is the World Wide Web. The tremendous increase in number of users in the WWW necessitated the use of caches that reduced the load on web servers. Therefore, it is possible to satisfy the higher demand caused by the increase in users. If this rise continues, additional caches can be installed to keep the systems scalable.
- *Geographical scalability*: The WWW is also a good example for *geographical scalability*. Information is offered from virtually any location in the world and users can access this information no matter where their location is. A distributed system is geographically scalable if effects caused by a wide geographical distribution of the system are hidden. A user who requests a web document and is served by a local cache receives the document with the same performance as if it had been requested from

a local server, although the original document might be located somewhere else.

- *Content scalability*: Content scalability is usually applied to multimedia content such as audio and video or even pictures. It characterizes content whose storage size and bandwidth requirements for transmission vary. MPEG-1 Audio Layer III (MP3) objects are a relevant example for this type of scalability. The standard [12] allows different types of encoding (the sampling frequency is varied) and, thus, the resulting MP3 files and the required bandwidth for transmission are of different sizes. In combination with multimedia objects, content scalability is also related to the quality of the object. An MP3 audio file generated with a higher sampling rate has a better quality than an MP3 file generated from the same original with a lower sampling rate.

Content scalability can contribute to the overall scalability of a distributed system. A specific example for content scalability in a distributed system is discussed in more detail in section 2.4.5.

2.2 REPLICATION

Replication is one technique for increasing scalability in a distributed system: '*Replication not only increases availability, but also helps to balance the load between components leading to better performance*' [9]. Therefore, a distributed system should make use of replication in order to increase its scalability.

One special form of replication is client-initiated replication [9] which is also called *caching*. It is distinguished from server-initiated replication by the fact that the decision to create a replica is made by the client of a resource and not by the owner. Scalability is increased in a similar way as with replication.

In the following sections, the terms replication and caching are discussed within the focus of video distribution systems. The distinction between replication and caching in the scope of video distribution systems is necessary because a certain class of content can make the application of one or the other technique more suitable. The distinction between replication and caching can also be observed in the WWW where content is actively distributed (replicated) or autonomously cached. For the first case mainly overlay networks like CDNs (content distribution networks) are used for the distribution, while in the second case no such overlay networks are necessary.

2.2.1 Server-initiated Replication

Replication is more efficient than caching in the case of a low *read-to-update* ratio. That is, it is very likely that few read requests are made to a replicated content until the replica must be updated. In addition, replication provides consistency by actively updating the replicated content. News is a popular example for content that should be replicated, i.e., actively pushed into local storage nodes. The content of a news video might change several times during a day and it might be requested by many users during a certain time interval. If, in addition, several news videos are created that are only of regional (geographical) relevance, the distribution system can determine to which storage nodes a specific news video should be pushed.

The main advantage of replication is load levelling which can be performed to reduce the origin server's load. For example, the popularity of a news video might be very high for a short duration. Without replication the origin server might not be able to handle all requests for that video. Pushing the video into local storage nodes reduces the possibility of overloading the origin server. Thus, applying the replication technique to news videos is efficient because the amount of network traffic is reduced. Additionally, it is scalable because server load is reduced and, thus, more clients can be supported, and the reliability of the system is increased. For example, in the case of a server failure, the news video can still be streamed from the local storage node, although, if the failure period exceeds a certain amount of time, consistency can no longer be guaranteed. It was shown, in this context, that the location of the replica (i.e., the storage node chosen to store the replica) has an important influence on the scalability of the distribution system and the availability of the content [13]. Other types of content that can be distributed by replication are all kind of objects

Table 2.1 Comparison of replication and caching

Replication	Caching
owner-initiated	client-initiated
low *read-to-update* ratio	high *read-to-update ratio*
increases system scalability	increases system scalability
increases reliability	increases reliability
suited for content that is of high popularity for a short duration, e.g., news	suited for content with lower or unknown popularity, e.g., on-line lectures

with a very high popularity, e.g., the latest blockbuster movie in a VoD system.

Replication has the drawback that the scalability of the system is constrained by the way updates for an object are performed. In contrast to caching, an additional mechanism is needed that keeps track of the existing replicas in a content distribution system in order to be able to update those replicas, if necessary. It might also occur that a replica is created on a storage node but is never requested by a client, thus leading to unnecessary resource consumption.

2.2.2 Client-initiated Caching

When the *read-to-update* ratio is relatively high or hard to predict, an approach that is based on the caching technique is more efficient. In this case, a copy of the original object is only created on the cache, if a client requests this object and the local caching policy on the cache decides that this object should be cached. This technique is efficient for objects which are of a lower popularity and are updated rarely or never. Objects that fit into this category are, for example, less popular movies, on-line lectures, or recorded sports events. The latter might be interesting for users who did not have the chance to take part in the live broadcast of the event. As in the case of replication, the amount of network traffic can be reduced if a cached object is requested more than once from the cache. In contrast to replication, caching has the advantage that objects are only distributed to a cache at which a request for this object has been made from a client. A distribution architecture that is based on caching increases the scalability of the system, since server load is reduced and more caches can be added should the number of users increase. Reliability can be increased with caching because the content of a cached object rarely changes. Therefore, server outages or link failures between server and cache can occur for a longer period of time and still clients can be served from the cache. Caching is not as constrained by the update procedure as is the case for replication because the server does not have to keep track of the caches that keep a copy of its content. The cache itself is responsible for the invalidation of cached content. Depending on the behaviour of the cache, a server might even not recognize that a copy of its content is stored on a cache. Yet, it is not clear if content-providers would allow autonomous caches in a distribution infrastructure. This decision is mainly based on copyright issues [14] which is not in the focus of this book.

2.3 VIDEO DISTRIBUTION SYSTEM TERMINOLOGY

Before the presentation of the specific video distribution architecture that was outlined in the scope of this book, the building blocks that are necessary to build video distribution systems are introduced in more detail. Since the terms to identify the elements that constitute a video distribution system are not consistently named in the literature, this introduction is necessary in order to avoid confusion. A possible configuration of a video distribution infrastructure is shown in Figure 2.1.

2.3.1 Origin Server

Origin servers store the original version of a video object. Those servers are in general controlled by content-providers. That means, the content-provider decides what content it offers and at which point in time new content is offered by actively storing this content on the origin server. The content-provider may also allow (or not allow) a cache to store a copy of the original content on its local storage. In the architecture presented here it is assumed that origin servers and proxy caches are owned by a single or cooperating CDN operators. Therefore, proxy caches can obtain copies of the original content without any restrictions. For reasons of simplicity the term *server* is used to denote an origin server in the remainder of this book.

In the case of a hierarchical distribution infrastructure (see section 2.3.5), the origin servers are located at the top level of the hierarchy.

2.3.2 Proxy Cache

A request for a specific video object is directed from the client to its nearest proxy cache. The proxy cache has then two options to deal with this request: (a) if the requested video object is already cached, it simply starts streaming this object from the cache to the client; (b) if not, it forwards the request to another proxy cache or an origin server. Content discovery mechanisms such as those provided by the Internet Cache Protocol (ICP) [15] or the Real-time Streaming Protocol (RTSP) [16] might be used to forward the client's request to a node of the distribution infrastructure that stores the requested object. Depending on the local caching strategy the proxy cache decides whether it caches the requested object or not. The outcome of the caching decision gives rise to two different scenarios:

- *Object is cached*: The cache informs the originator of the content that a local copy of the requested object is created on the cache. Based on the distribution mechanisms the sender of the stream might either set up a multicast stream, which is joined by proxy cache and client, or a unicast stream that is forwarded through the proxy cache to the client (see Figure 2.1). The issue of how a reliable transport between server and cache can be achieved is addressed in section 3.6 and Appendix A.

- *Object is not cached*: In this case, there is no general need for sending the stream to the cache. Simply streaming the data from the server to the client is sufficient.

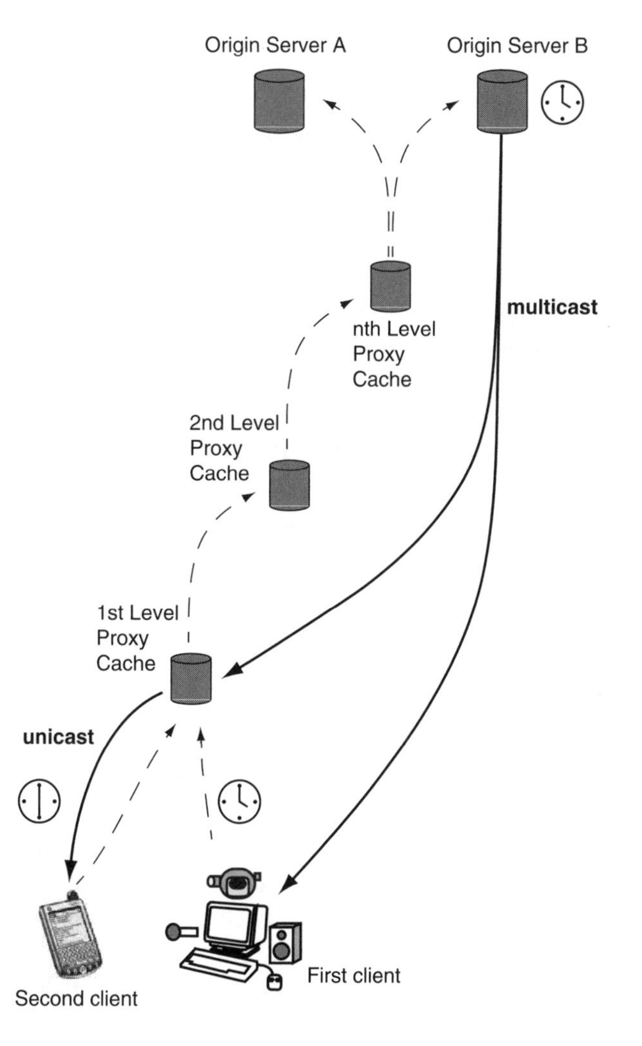

Figure 2.1 Caching hierarchy.

In addition to the functionality that must be provided by a proxy cache (e.g., cache replacement, extended signalling, etc.), it also offers the same functionality as an origin server. Thus, in terms of provided functionality, a proxy cache can always be seen as an extension of the origin server.

In the literature *proxy cache* and *cache* are used interchangeably, while always this definition of a proxy cache is meant. Throughout the remainder of this book the term *cache* is used.

2.3.3 Cache Replacement

Since the available storage space on the cache is limited, cache replacement has to be performed on a cache to increase the efficiency of a cache. Cache replacement is not limited to video caches but is also applied in all kinds of caches, such as memory caches or web caches [17]. Cache replacement mechanisms were originally developed for memory cache. With the advent of web caches, replacement strategies for those caches became an interesting topic. Unfortunately, these strategies cannot be applied directly to video caches since the characteristics of the objects that are cached are different. Video objects usually have a high read-to-update ratio compared to typical web objects, such as web pages. In addition, transfer times are much higher and the ratio of cache size to object size is much smaller. In the case of a web cache and a video cache with the same storage space, the first can store a larger number of objects than the latter owing to the difference in object size. An introduction to cache replacement for video caches is given in reference [4].

2.3.4 Client

Throughout this book, it is assumed that all clients are connected to the Internet. This is achieved via different access technologies, such as Local Area Networks (LANs), ADSL, cable modems, wireless networks, or modems. The clients can be all kinds of devices that are able to deal with multimedia data (in this specific case, video streams). For example, clients are set-top boxes, standard PCs, PDAs, and mobile phones. This implies a very heterogeneous environment concerning characteristics such as access bandwidth, computing power and display capabilities of the client. Each client obtains the information to which cache its requests for a video object should be directed. This can be done manually by setup parameters or, if the object is requested via a web server, through additional HTTP [18] information.

2.3.5 Logical Overlay

Another important issue is the placement and interconnection between servers, caches, and clients. One traditional approach is a hierarchical distribution approach as shown in Figure 2.2. In this case, the servers are located at the top level of the hierarchy while caches are located in the intermediate levels and clients always on the lowest hierarchy level. Client requests are directed to caches of the lowest hierarchy level and only forwarded to caches in the next higher level, if necessary. The distribution hierarchy can consist of a different number of levels with a minimum of at least two levels which would lead to a video distribution system only consisting of servers and clients. Another approach is often described as *cooperative caching* [19] in which the forwarding of requests is not as restricted as in the case of a hierarchical distribution infrastructure. With this type of infrastructure caches can cooperate with each other independently of their location in the overlay. Thus, requests can be forwarded to any cache or even server in the infrastructure. A comparison between the two types of distribution infrastructure is shown in Figure 2.2. The links between the single entities of the infrastructure represent the possible path of request messages and do not reflect the physical layout of the infrastructure.

The new transport mechanisms presented in this book are suited to both kinds of hierarchies. For reasons of simplicity, further examples and measurements have only been made on the basis of a hierarchical infrastructure. This does not restrict the presented transport mechanisms, since the performed investigations and design decisions for these mechanisms are not limited to a certain type of distribution infrastructure, as is shown throughout this book. Nevertheless, cache replacement mechanisms can depend on the logical overlay. For example, with cooperative caching a new object must

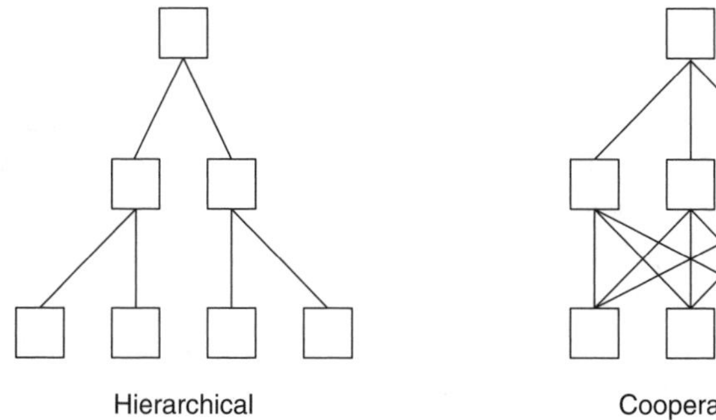

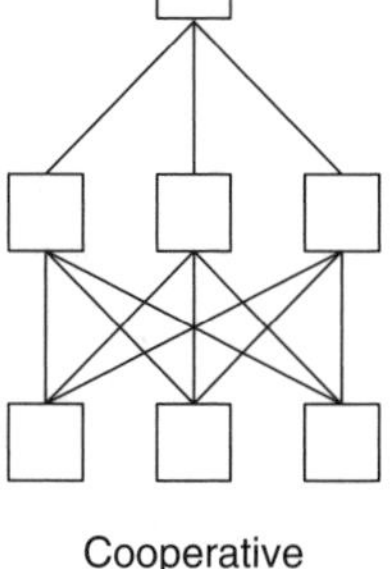

Figure 2.2 Logical overlay.

not necessarily be delivered from a cache or server that is located in a higher hierarchy level. The cache replacement presented in Chapter 6 is independent of the distribution infrastructure, since only local decisions about content that should be discarded from the cache are made.

2.3.6 Video Object

Any video content that can be streamed via the Internet is defined as a video object. An example of a video object is an MPEG-1 [20] encoded movie that has a duration of approximately 90 minutes. No restrictions on the encoding format, the duration of the actual content of the video object are made. Thus video objects could also be short music video clips or lectures as they are provided in distributed learning systems such as *eTeach* and *BIBS* [21].

2.3.7 Video on Demand (VoD)

The term VoD is widely used for systems that allow one to watch a certain video content at any point in time via communication systems such as cable TV, satellite or the Internet. VoD is used to describe on the one hand services such as pay-per-view that allow viewers to watch a movie on a digital TV channel (either cable or satellite) with the restriction that those movies are only offered at certain points in time; on the other hand VoD describes services such as True-VoD (TVoD) that allow a client to watch a video content immediately after its request. According to reference [3] there are five different types of services, classified on the level of interactivity allowed:

- *Broadcast (NoVoD)*: There is no interactivity in NoVoD. One example of this type of VoD is the traditional TV channel.
- *Pay-per-view (PPV)*: Viewers sign up and pay for a specific program but still cannot influence the way, when or how the content is shown.
- *Quasi VoD (QVoD)*: Viewers can be grouped together based on the threshold of interest. For example, viewers interested in sports events are joined in a group at which soccer games are broadcast. The viewer has rudimentary control in having the ability to change group and, in this way, receive a different type of content; but interactivity is not supported.
- *Near VoD (NVoD)*: Interactivity such as fast-forward and reverse is only allowed in certain time intervals. Even when a viewer decides to start watching the video it might take a certain amount of time until the content is displayed. One way to realize NVoD is to broadcast the same content on different broadcast channels while the start of each broadcast is shifted in time. For example, a movie with a total length of 60 minutes is broadcast

over 12 channels with a time shift of 5 minutes. In this scenario, a viewer might have to wait up to a maximum of 5 minutes until the playout starts, and the granularity for fast-forward and rewind is limited to 5-minute segments, as shown in Figure 2.3. The limitation to 5 minutes for fast-forward and rewind results from the fact that these operations are realized by switching to a different channel.

- *True VoD (TVoD)*: In this case, the viewer has complete control over the session presentation. The viewer can jump to any position and perform operations similar to those that are offered by a VCR. In the case of TVoD, one channel cannot be shared by several viewers. Of all the five types, TVoD is the most challenging, since for each single request a new channel (i.e., a new stream) must be set up and it is not guaranteed that the video object is streamed linearly. The viewer might perform nonlinear operations such as fast-forward, fast-rewind or pause.

In the scope of the work presented in this book the Internet is regarded as the underlying communication system for VoD, with the focus on True VoD. The decision to focus on TVoD was driven by the fact that it offers the best possible service in terms of interactivity and, therefore, increases the attractiveness of VoD on the Internet compared to other services which are already available via traditional broadcast media (e.g., PPV). In section 2.6, an application is presented where the level of interactivity is increased by the offered TVoD service.

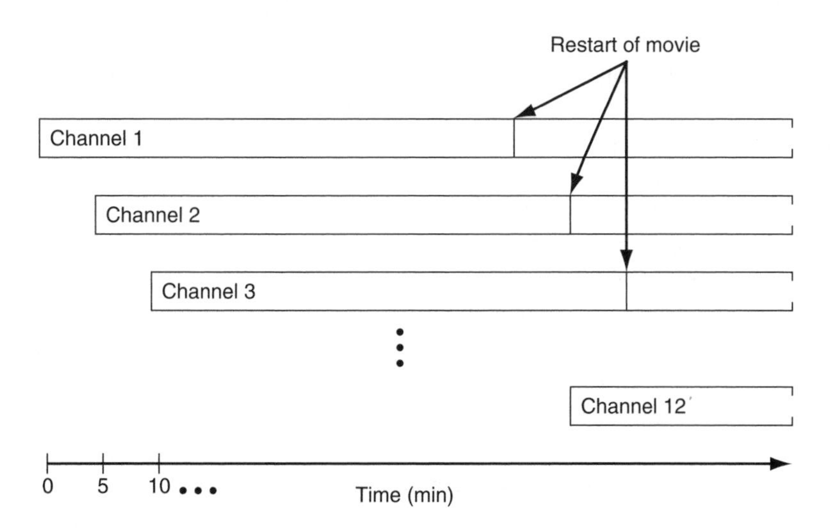

Figure 2.3 Near VoD.

2.4 ARCHITECTURE

2.4.1 A Snapshot of Today's Internet Infrastructure

After the identification of the elements that form a video distribution infrastructure, the *scalable adaptive streaming* (SAS) architecture is presented in this section.

The main goal of this architecture is to support the efficient distribution and streaming of video data in today's Internet. Most of the existing approaches require techniques that are not fully established in the Internet so far, or may probably never be established. One example is the work that has been performed on broadcast and multicast schemes in order to reduce load on video servers (see section 3.6). All of the existing approaches assume that the bandwidth between client and server is sufficient and no congestion on those links occurs. Therefore, those approaches can only be reasonably applied in environments where bandwidth reservation between the different entities of the distribution system can be performed, as, for example, with open-loop control mechanisms such as RSVP [22]. Unfortunately, those mechanisms are not deployed in the Internet and it is rather debatable if those mechanisms will ever be used in the global Internet. *Prefix caching* (see section 3.6) is an exception that assumes that the bandwidth between server and cache might not be sufficient. It has the drawback that only (temporal) parts of a video are cached and, thus, in the case of a link or server failure, the streaming of the video cannot be continued.

The architecture presented in this book is based on the techniques that are available in the Internet of today but is also designed in a way that allows adaptation to possibly new mechanisms and techniques in a future Internet.

Although there has been an immense amount of work on quality of service (QoS) [23], the only service that is currently offered in the Internet is best-effort. That means, resources in the Internet cannot be reserved and, thus, no guarantees for a certain service (e.g., a guaranteed amount of bandwidth, packet loss, or latency on the link between two nodes) can be offered. Since traditional streaming approaches make use of UDP as the underlying transport protocol, one might naively assume the lack of service guarantees in the Internet less problematic than would be the case for TCP. Yet, the lack of a congestion control mechanism in UDP leads to an unfair behaviour against TCP traffic. So far, UDP traffic in the Internet makes up only a small fraction of the total amount of traffic. This could change in the near future if, for example, streaming applications become more popular, leading to a situation in which the performance of TCP-based applications is heavily affected by non-congestion-controlled traffic. Several investigations on streaming media in the Internet have shown that VoD is becoming more popular, and it is very likely that this increase in popularity will continue in the near future. For example, a study in 1997 [24] did not report any remarkable amount of streaming media, while a study in 1999 [25] reported that 14% of the total amount of transferred web traffic belonged to audio and video content [26]. This trend is confirmed by a more recent investigation [27].

One solution to this problem of TCP-unfairness is to perform streaming also in a congestion controlled manner. In recent years, there have been several proposals on such TCP-friendly mechanisms [28] which are presented in section 3.4.

Recent developments in the end-system market have also increased the heterogeneity of end-systems and access links that are used by those systems. In the beginning of the Internet, end-systems were mostly located in local area networks built on Ethernet or Token-Ring technology; thus, access networks were mostly homogeneous. With the increasing popularity of the Internet, brought about by the success of the WWW, and new access technologies such as wireless LAN (WLAN) or DSL, the situation has changed. Future scenarios might even include all-IP based wide-area wireless networks, as is discussed for fourth-generation wireless networks. Therefore, a video distribution architecture must be able not only to provide mechanisms for TCP-friendly streaming, but also to allow adaptation to a wide spectrum of access link capacity and end-system characteristics.

In the following sections the benefits of using caches in a video distribution infrastructure and two classes of scalability on which the SAS architecture is based are presented. A scalable adaptive streaming architecture allows one to perform video distribution in the Internet of today that is characterized by best-effort service and a wide heterogeneity regarding end-systems and access networks.

2.4.2 Advantages of Caching

Studies on access patterns and file characteristics of video objects [29, 30, 26] have shown that caching can be advantageous for the following reasons:

- Most video applications follow the write-once-read-many principle; hence, cache consistency is not a major issue.
- Access to videos exhibits a strong temporal locality. That means, it is highly probable that an object that was accessed recently with a high frequency will be requested again soon.

- Broadcast (see reference [31] for an overview) and multicast mechanisms (e.g., [32–35]) used to reduce the load of the original server can only be used for QVoD or NVoD, but not for TVoD.

Additionally, caches for video objects share the well-known benefits introduced by web caches:

- The fault tolerance of the system is increased since a failure of the server does not necessarily lead to a service interruption for the client, if video objects are cached completely.
- Caches store content closer to the user and, thus, reduce start-up latency and network load.
- The load generated at the server is reduced and distributed over several caches.

Yet, in contrast to web caches the characteristics of the data to be stored are very different. High-quality video files are much larger than most web pages and, therefore, different caching strategies are used in caches for VoD systems [4]. The distribution process for video files is further complicated by the fact that the transmission is much more time- and bandwidth-consuming. Finally, the popularity models for video caches are different to those for web caches [5, 26]. Nevertheless, investigations on the popularity of video objects have shown that approximately 80% of the user requests are concentrated on 20% of the total amount of available videos [36, 37, 5]. These results indicates that caches can be a very effective means in relation to scalability and fault tolerance in a video distribution system.

The mechanisms for video distribution presented in this book do not have the goal of optimizing video distribution by investigating several caching concepts (hierarchical, cooperative) or of investigating cache replacement strategies, as was done in reference [4]. The focus here is on transport issues related to TCP-friendly streaming and caching which can be used by either hierarchical or cooperative caching and does not affect cache replacement strategies. With the aforementioned benefits of caches in video distribution systems in mind, the SAS architecture was designed with caches being one major building block of the infrastructure.

2.4.3 VoD without Scalable Adaptive Streaming

Before introducing the SAS architecture a short description of a video distribution infrastructure that is not supported by SAS is given. This description is used to clarify the differences between a traditional and an SAS-based distribution infrastructure.

Without SAS, video is streamed via standard UDP which does not allow an adaptation to the actual network conditions on the route between server and client. In addition, the content offered in such a VoD system is not scalable. This results in two possible scenarios, depending on the available bandwidth on the links between server and client. As long as the bandwidth on the links is sufficient the service can be offered to the viewer in full quality. In the case of insufficient bandwidth the service quality decreases to a point where it is unacceptable for the viewer to watch the video. Owing to uncontrolled data losses the resulting quality of the presented video is either very bad or, even worse, complete frames or sequences might not be rendered because of missing data (see Figure 2.4). Thus, such a VoD service can be seen as a *binary* service: either the quality of the client is at its best or it becomes very bad. Any steps in between do not really exist. It is obvious that the resulting quality depends on the amount of lost data, but it cannot be controlled which of the data is lost during the transmission. Investigations on MPEG-1 video sequences [38] have shown that selective losses of only 1% of the total amount of data can decrease the quality to a level where the content is no longer recognizable. An investigation by Boyce and Gaglianello [39], which measured the losses of UDP-based streaming

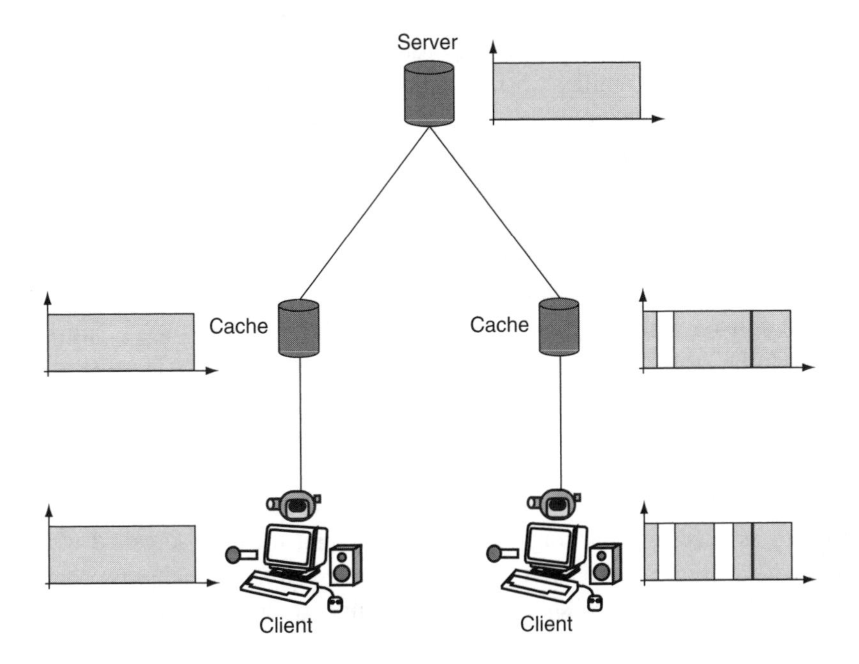

Figure 2.4 VoD without scalable adaptive streaming.

sessions in the public Internet, has shown that loss rates as low as 3% can lead to frame error rates as high as 30%.

An additional problem is the support of clients such as mobile devices which may not have sufficient computing power to render and display the content if offered in a non-scalable format.

Another approach to avoiding these uncontrolled losses is to combine a TCP-based transport with a sufficient buffer at the client. This solution has the drawback that the start-up latency can be high (the buffer has to be filled initially) and that re-buffering may occur in the case of a buffer underrun. Both effects can be quite annoying for the viewer.

2.4.4 System Scalability

A distribution system based on SAS combines *user* and *geographical* scalability (see section 2.1) in what is referred to as *system* scalability. The decision to combine both scalability concepts into one is prompted by the fact that the introduction of caches affects both, as is shown in the following.

As with traditional web caches, caches for TVoD systems allow one to store content closer to users, reduce server and network load and increase the system's fault tolerance. It is important to mention that system scalability here means the scalability of the whole video distribution system, in contrast to approaches that increase the scalability of a single component. For example, an enormous amount of work has been performed on the issue of single-server scalability. Although analysis of media streaming workload has shown that the application of techniques such as *patching* [40] or *bandwidth skimming* [41] are justified, they are not sufficient. First of all, the fault tolerance of the system is not increased, because without caches the server would still be a single point of failure. In addition, video objects are less likely to be stored near the client if no caches are involved. Thus, start-up latency might be increased and quality degradation of the stream might occur because the potential of a link failure or congestion is more likely on a WAN connection.

Figure 2.5 depicts an example for system scalability where hierarchical caching is used as a caching concept. As caching method, so-called write-through caching[†] is employed, where a requested stream is either forwarded through the cache or is streamed via a multicast group which client and

[†] Adopted terminology from memory hierarchies.

caches join, if the cache replacement strategy decides to store the requested video on the cache.

Subsequent clients can then be served from the cache (see Figure 2.5). This technique has a lower overall network load in a TVoD system than a method where the video is transported to the cache in a separate stream using a reliable transmission protocol (e.g., TCP) [42]. On the other hand, write-through caching requires a reliable (multicast) protocol to recover from packet losses. The design and implementation of such a protocol, called Loss Collection RTP (LC-RTP), which fits particularly well in a TVoD architecture and employs write-through caching, is presented in Appendix A. As can be seen in Figure 2.1, the SAS architecture can increase fault tolerance in a video distribution system, especially, if we keep in mind the temporal locality characteristics that were observed for video traffic in the Internet. Caching complete video objects would allow one to still serve a large number of clients despite a server or network (between server and cache) failure.

In addition both the usage of caches and write-through caching reduce the server and network loads and, thus, increase the number of clients that can be supported by the system. In the next section, content scalability, which allows congestion control for streaming applications, is introduced.

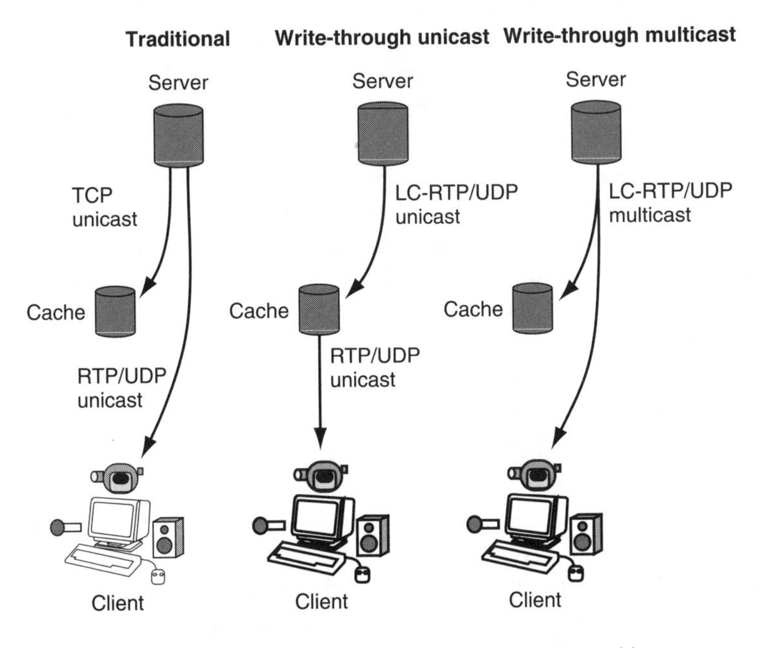

Figure 2.5 Comparison of transport methods for caching.

2.4.5 Content Scalability

In contrast to elastic applications whose traffic can be spread over time in order to adapt the transmission rate to the available bandwidth on the network, quality adaptation is needed to enable congestion control for inelastic applications such as streaming.

For elastic applications such as FTP, data may not arrive at the receiver at a certain point in time. For example, in the case of an email transmission all the data of the email is transmitted to the client application and after that the user can open the email for reading. This is different from inelastic applications where the delivery of the data is time-critical. For example, in the case of a MPEG-1 video that is streamed to the client, certain parts of the video must arrive at the client at specific deadlines to allow correct representation at the client.

However, quality adaptation does not solely serve congestion control purposes but also satisfies the needs of the great variety of heterogeneous clients that exist in the Internet. Even in cases where, owing to the provision of network QoS mechanisms in the Internet, congestion control must not be performed, clients might be connected via access networks that have different bandwidth characteristics. Thus, clients are able to receive the requested content with different rates. A client which is directly connected to an Ethernet LAN might be able to receive an object with a rate of several megabytes per second, while a client connected to XDSL might be limited to receive an object with a rate of several hundred kilobytes per second. This could not be realized if the content offered on the server were only a monolithic encoded file, such as an MPEG-1 file.

In today's video distribution systems the heterogeneity problem is solved by offering only low-bandwidth streams, thus punishing clients which could receive higher bandwidth streams leading to a better perceived quality of the video. Investigations on RealVideo over the Internet [43, 44] state this assumption, since none of the requested video objects had a bandwidth higher than 500 kbps.

Layer-encoded video, i.e., video that is encoded in two or more layers, represents a suitable method to allow for this quality adaptation. Besides the layer-encoded format there are other alternatives such as adaptive encoding or switching between different encodings of one original content [45] (also described as *dynamic stream switching (DSS)* or *simulcast*). If both techniques (DSS and layer-encoded video) are compared with each other, it becomes obvious that layer-encoded video has the advantage that it can adapt to network conditions in a finer granular manner. With DSS that would require a larger number of different encodings (regarding the transmission bandwidth) of one original content to be stored on the server or the cache. For example, to offer the same quality levels with DSS as a layer-encoded video consisting of three layers, three independent video objects, each encoded with a different rate, must be available. In addition, the switching between objects of different rates at the client or the cache cannot be performed as easily as in the case of layer-encoded video. For example, in the case of a video that is encoded in several MPEG-1 bitrates (alternatives) a switching between these alternatives can only be performed at the next I-frame. Switching on a random position in the video is not possible, owing to the intra-coding characteristic of MPEG-1 [1].

Figure 2.6 shows the advantages of scalable and non-scalable encoding formats for the case where clients in the distribution infrastructure are connected to the Internet via diverse access technologies offering different maximum transmission rates. As mentioned earlier, in the case of non-scalable encoding formats one or other class of clients may experience rather limited service. In the example shown here the clients connected to the high-bandwidth access network get a low-quality service, since they can only receive the low-quality stream. If only one quality version of the content can be offered, content providers have to apply this principle in order to make sure that the clients connected to low-bandwidth networks have a chance to use the service at all. Were the video to be encoded in a bandwidth higher than the actual bandwidth, there would be no possibility of these clients to receiving this video in acceptable quality. This problem can be circumvented by applying layer-encoded videos. In this example two different techniques of layer-encoding are shown. The first one is a discrete layer-encoded technique, while the second one allows much finer adaptation to the available

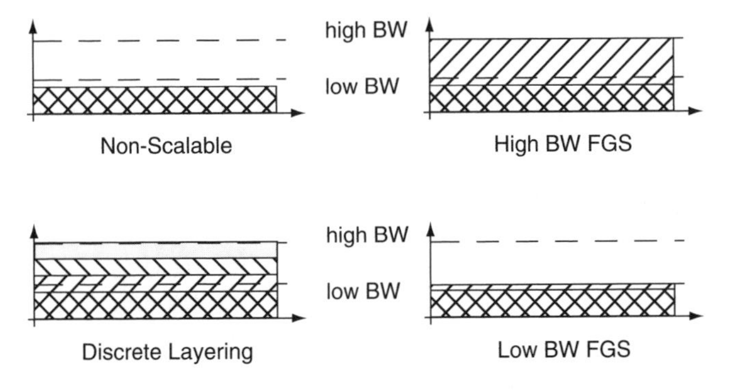

Figure 2.6 Adaptation to access network bandwidth.

bandwidth. Nevertheless, whichever of the two techniques is applied the benefit compared to non-scalable encoding is obvious.

For the remainder of this chapter the discrete layer-encoding approach will be considered. A more detailed overview of existing layer-encoding techniques is given in section 3.3.

2.4.6 Combining System and Content Scalability

As already mentioned, TCP-friendly streaming is a major issue in SAS. There have been several proposals on how to achieve TCP-friendly congestion control using hierarchically layer-encoded video transmissions, e.g., [46], [47], or [48]. Existing approaches for TCP-friendly streaming are presented in section 3.4.

Figure 2.7 (a) and (b) show two possible versions (layered and DSS) of a cached video if a TCP-friendly video transmission is combined with write-through caching. Obviously, in both cases, the cached copy of the video exhibits a potentially large number of missing segments from different layers

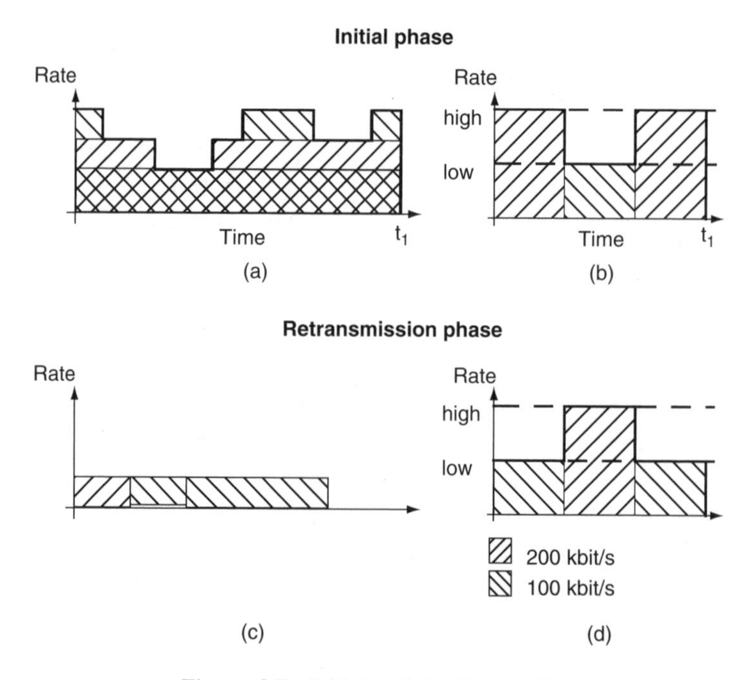

Figure 2.7 Initial cached video quality.

(or complete encodings as in the case of DSS). Note that the exact shape of a cached video content is a function of the congestion control mechanism being used by the TCP-friendly protocol.

Clients that request the same content at a later point in time and would be served from the cache do not have the chance to receive the video in full quality if no extra measures are taken. Thus, a mechanism is required that improves the quality of the cached content. Figure 2.7 (c) and (d) depict the parts that would be identified by such a mechanism and be transmitted from the server to the cache leading to a full-quality copy of the video object. The benefit of using layer-encoded video instead of DSS becomes even more clear if we compare the amount of data that has to be additionally transmitted for both encoding techniques. In the case of DSS, complete parts of an encoded video must be transmitted instead of only missing segments of certain layers as is the case for layer-encoded video (see Figure 2.7 (c) and (d)). Thus, with layer-encoded video less network resources and storage space at server and cache is consumed for the transmission of missing segments.

Owing to the distinct advantages of layer-encoded video compared to the other methods that allow adaptive streaming presented here and in section 2.4.5, only layer-encoded video is regarded in the remainder of this book.

In the following, transmissions of missing segments from the server to the cache caused by the cache's request for these segments are called retransmissions. At first, this definition might be confusing because some of the missing segments may never have been transmitted at all (due to congestion). On the other hand, the cache cannot distinguish between packets which were not sent at all and dropped packets (e.g., queue overflow on one of the intermediate routers). Thus, for the cache every packet that was not transmitted initially appears as a retransmitted packet.

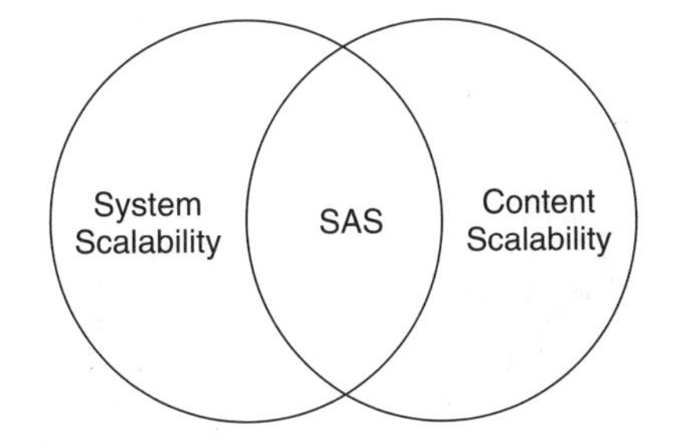

Figure 2.8 Combination of system and content scalability.

2.4.7 VoD with Scalable Adaptive Streaming Support

Figure 2.9 shows the same basic architecture as in Figure 2.4 with additional scalable adaptive streaming support. In contrast to the binary service, as described in section 2.4.3, this new video distribution architecture allows more than two quality steps. This is mainly caused by the fact that the multimedia objects (e.g., videos) are encoded in a scalable format and, in addition, the streaming is performed adaptively to the conditions in the network. As shown in Figure 2.9 video distribution with SAS support makes it possible to deliver a video stream in more than two quality steps to the client. In addition, the scalable content is also well suited to supporting a large variety of clients (PCs, PDAs, etc.). This book presents new mechanisms, which are part of the SAS architecture, that also try to deliver the video in the best possible quality to the client. The two sections in Figure 2.9 which are marked by a circle give a simple example that demonstrates how the quality improvement can be achieved. The initially cached version of the video misses some segments and, thus, would lead to a lot of quality variations if the video were to be streamed to the client as it is stored on the cache. With the new mechanisms presented in this book, the amount of

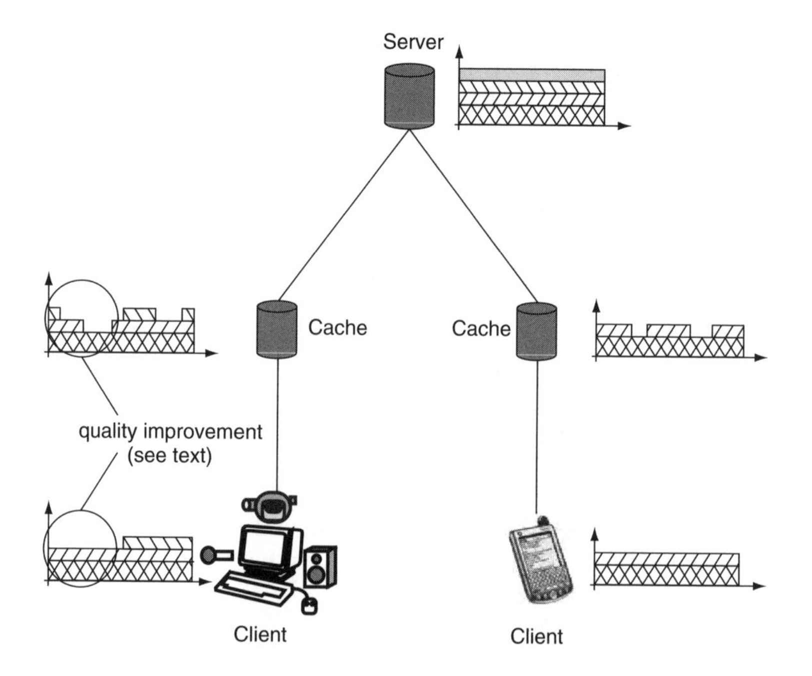

Figure 2.9 VoD supported by scalable adaptive streaming.

quality variation can be reduced, as is shown for the video that is finally streamed to the client. The mechanisms used to reduce the amount of quality variation are presented in Chapter 5 and Chapter 6.

2.5 SCENARIO FOR SCALABLE ADAPTIVE STREAMING

To clarify why it is important to combine both kinds of scalability in a video distribution system that suits well for today's and the future Internet, a scenario that uses the SAS architecture is presented in the following section. The goal is to present how a video distribution infrastructure can benefit from the integration of system and content scalability. The example tries to reflect a typical scenario such as can occur in today's Internet.

The scenario shown in Figure 2.10 depicts a heterogeneous distribution system consisting of two subnets that are connected to the Internet backbone. In each of these subnets a cache is located to which all client requests are directed. Subnet A has a wireless infrastructure in which only homogeneous clients (in terms of link bandwidth) are connected, while Subnet B has a heterogeneous infrastructure. In the case of Subnet A it becomes clear that the maximum of two layers is sufficient for all clients, since all clients are connected via a homogeneous access network. It is assumed that the original videos stored at the server consist of four layers, as shown in Figure 2.10.

In Subnet B, the content might be first requested by a PDA (client 2) in a lower quality because of its restricted access bandwidth. A subsequent client requesting this content might be a high-end PC which would like to receive the content in a better quality and has an access link with high-bandwidth characteristics. Its capabilities allow for additional transmissions from the server to the cache in order to improve the quality of the cached content. Depending on the order client 1 and client 2 are requesting the video, it might occur that client 1 receives a stream from the cache that has reduced quality because no higher quality is available at the cache. This can be for two reasons: either the stream was originally requested by client 2 which did not request more than two layers or, because of space constraints, the cache replacement strategy might have dropped the two upper layers. If a retransmission mechanism is used, this situation could be circumvented. After initially delivering the stream to the client 2 the cache starts to request missing segments of certain layers or even complete layers upon a request from client 1. This would be the case for segments from the second and third layer, as shown in the scenario in Figure 2.10.

In some cases it may not be useful to cache all layers of a video, as was already mentioned for the case of Subnet A. It is assumed that a certain number of video objects are already stored on the cache and none of those videos was requested in a higher quality than two layers by the clients which are served from this cache. Now the cache can presume that none of its clients can receive a video in a better quality and for the caching process of a new video no more than two layers would be cached. In addition, the clients can also give hints about the maximum rate they are able to receive data with (as shown in Chapter 8). In the retransmission phase only missing segments of these two layers would be requested. In addition, this information could be used for the cache replacement strategy. Storage space needed for the caching of new content could be gained by deleting all layers above the second layer instead of deleting complete videos. Applying such a mechanism can lead to a two-dimensional cache replacement where one dimension is the quality (number of layers) and the other dimension is the time (e.g. as in prefix caching (see section 3.6)).

2.6 AN EXAMPLE APPLICATION FOR SCALABLE ADAPTIVE STREAMING

In this section, an example application that could benefit from the incorporation of the SAS video distribution architecture is given. The goal of this section is to explain the application of the SAS video distribution architecture in the Internet. This example is thought to give the reader more detailed information about the context of the work presented in this book. Also, the application described below is not the only application that could benefit from the SAS architecture. Other possible applications are TVoD service that offers a large variety of video objects having a large variance in their popularity (e.g., not only the latest blockbuster movies but also classics are offered) or business TV of a large company that has locations in different regions, countries or even continents. In this section, distance learning is the application scenario that is used to demonstrate an application of the SAS architecture.

In Europe, it is very common for students to spend a certain amount of time at another European university as part of their studies. Unfortunately, not all students get the opportunity to attend the official European exchange programme (ERASMUS), for many reasons. Offering on-line courses would allow students to study with loosened time and location constraints. A person having a regular job could attend an on-line course at a university because

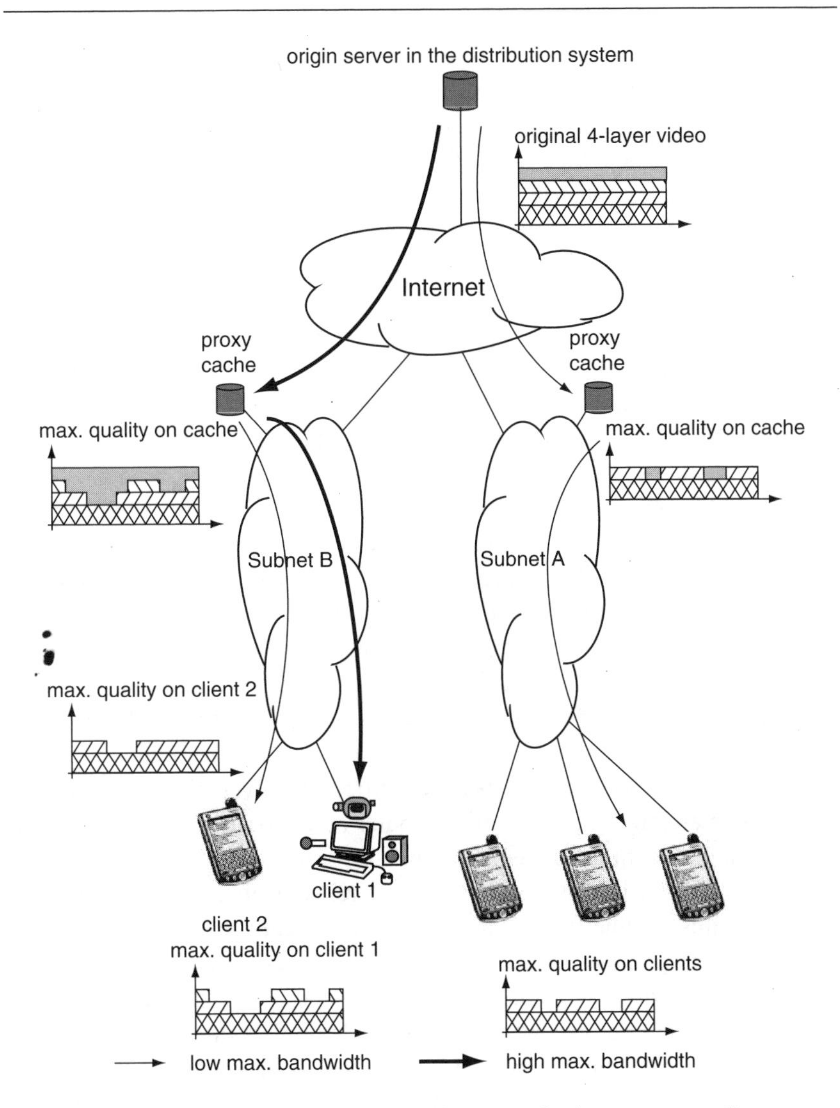

Figure 2.10 Scalable video distribution system for heterogeneous clients.

there is no necessity to attend the regular courses during the daytime. One way to offer on-line courses is to record the regular course that is given on a regular (mostly weekly) basis and make this course available for all subscribing students. There is also an advantage for the lecturers who could record a course in advance, having in mind that at the regular date another important event might happen (e.g., a conference the lecturer wants to attend). The course can be made available on-line at its scheduled date and every student has the chance to 'attend' it. Not only broadcasting the course (e.g., via multicast) as a live stream but also making it available for on-demand requests bears the advantage that students can 'attend' the on-line course when it fits into their time schedule. Another advantage of this method is that on-line courses could be shared between universities. For example, two universities might agree to offer on-line courses for all students of both. An example of on-line course sharing between universities is the German ULI project [49].

Another example that simply offers courses on-line is the MIT Open Courseware project [50].

These on-line courses would offer students a broader variety of courses and allow them to develop their language skills, if, for example, lectures from universities located at other countries are provided. Yet, arranging time schedules between universities might be difficult or even impossible. This could lead to the situation that there might be an interesting on-line course at a remote university that is given at exactly the same time a local lecture is given which the student wants to attend. Thus, the student would not be able to attend the on-line course if it is only distributed as a live stream. The need for on-demand, on-line courses is even higher if one imagines that such courses are offered across continents (e.g. Europe and North America) because of the different time zones involved. Also the beginning and end of a semester are not identical between different universities. Some of the on-line courses might become quite popular at a certain university because students might recommend them to each other or the lecturer is a well-known person in a specific research area. Thus, several students from one university will watch this on-line course, but it is very unlikely that all of them will watch it at one point in time. Measurements of user access behaviour in a distance education application [30] have revealed a high temporal locality. That means, several requests for the same video object often occur within a short time span. Caching the on-line courses at a cache located in the campus network of the local university would be beneficial in this case because:

(1) Load on the origin server at the remote site is reduced since, after caching the lecture at the local site, subsequent requests are not directed to the origin server.
(2) Fault tolerance is increased because students may still be able to watch the lecture even though the original server or the path to it has failed.
(3) Overall network load is reduced since only one complete lecture has to be streamed across the Internet backbone. In all other cases only local resources are consumed.

(4) Once the file is cached the need for an update is highly improbable; even if any specific lecture is given at the very beginning of the next semester, by then the video object representing the lecture should have long been discarded from the cache.

Even students at one university might use different access technologies to connect to the campus network and the Internet. For example, students in dormitories might be connected directly to the campus network via LAN technologies such as Ethernet, while some universities already offer wireless access on campus (e.g., WLAN) to their students. Finally, students might use ADSL, cable or traditional modems to get access to the campus network. This is very similar to the scenario as shown for Subnet B in Figure 2.10. An additional important fact is that students may be unwilling to pay for a certain service class that would ensure QoS guarantees on the link between the server, the cache and the client. Thus, an architecture for the distribution of on-line courses across universities should be well-suited for the best-effort service class, which is so far the only available class. This example also shows that such a system is not outdated if other service classes become available in the future Internet. Although service guarantees may be offered in the future, students might still be connected via access networks that do not allow the reception of the video stream in full quality.

3

Towards a Scalable Adaptive Streaming Architecture

This chapter focuses on work that is concerned with the building blocks of an SAS architecture. In recent years, a lot of effort was put in the issue of single-server scalability, which includes topics such as disk access, memory management and techniques that increase the delivery capacity by efficiently using network resources. These topics are very important in relation to performance for single entities such as servers or caches in the SAS architecture. They are not in the main focus of this book, since here the main focus is on the overall aspect of the system architecture and the interplay between its single components. For further information on single-server performance the interested reader is referred to reference [31].

The first part (section 3.1) of this chapter is concerned with available products in the area of video distribution systems and shows the shortcomings of these products in relation to the requests for an SAS architecture, which were defined in Chapter 2. The second part (section 3.2) regards the related work of standardization bodies which mainly influence standardization in the Internet and the digital broadcasting area. The work on video distribution has gained a lot of attention, which is shown by the fact that products have become available and that the IETF has inaugurated a new working group for content distribution (Content Distribution Internetworking (CDI)). Nevertheless, both available products and the work in the IETF mainly aim at the goal of simply moving content closer to the receiver, increasing fault tolerance and reducing server load.

Existing work on content scalability, which in the case of SAS is realized by scalable encoded video, is presented in section 3.3. Three of the most popular layer encoding schemes for video are presented and compared with each other.

Scalable Video on Demand: Adaptive Internet-based Distribution M. Zink

Related work on TCP-friendly streaming which allows adaptive streaming in combination with scalable encoded video, is presented in section 3.4. TCP-friendly streaming can be separated into two main approaches, window-based and rate-based. Section 3.5 shows existing approaches to adaptive streaming, where no caches are involved, so far. The existing work on caches for video streaming is very broad. The most important areas of this research work are presented in section 3.6. It shows the work on partial caching which can be subdivided into time-based and bandwidth-based caching. This work is relevant for SAS since layer-encoded video can be cached partially, both in the time-based and bandwidth-based domain. Reliable transport into caches allows the creation of identical copies of the original video on the cache. Section 3.7 gives an introduction to this topic and presents some of the existing work in the area. An overview on cache clusters, which is an additional approach to increasing the scalability of streaming systems, is presented in section 3.8.

3.1 PRODUCTS

Consequent upon the increasing popularity of video streaming in the Internet more and more products are available that offer streaming applications. These products can be divided into three major categories:

- server and client application;
- server, client and cache application;
- cache application only.

The first category is not of much interest, since it does not offer mechanisms for scalability as they are included in the SAS architecture. One example of the second category are the products offered by Real Networks [51] which offer a complete solution for video distribution in the Internet. There is very little information available about Real's products but this information reveals that neither TCP-friendly congestion control nor caching in the traditional sense is performed. In the case of a cache miss, the data is streamed directly to the client and a second, reliable (TCP-based) connection is established to the cache that is used to transport the requested video to the cache. Although streaming can be performed in an adaptive manner by using a flavour of DSS, called *SureStream*, the mechanism to adapt is based on RTCP feedback information and, thus, can hardly be considered as congestion control. The restrictions on the amount of RTCP feedback information that can be sent during a certain period is restricted. Therefore, the adaptation to changing conditions in the network takes longer than with TCP or a TCP-friendly mechanism.

The third category is represented by products from Kassena [52], Network Appliance [53], Inktomi [54], Blue Coat [55], Infolibria [56]. Novell Volera Media Exelerator [57], and CERTEON Media Mall [58]. Unfortunately, there is no technical information available about these products, except in the case of Kassena's MediaBase XMP which reveals that they make use of prefix caching [59] in their cache.

Not included in these three categories, because they rather make use of the products offered in the third category, are CDNs such as Akamai [60] or Cable & Wireless [61]. Usually, replication is applied in these CDNs as a mechanism to distribute the content into caches.

3.2 STANDARDIZATION

In relation to the Internet the most important standardization body is the Internet Engineering Task Force (IETF). Virtually every open standard protocol that is used in today's Internet has been standardized by the IETF. Also most of the available video streaming and distribution applications support IETF standards.

In parallel with the IETF, standards for video distribution have been developed by the Digital Video Broadcasting Project (DVB) and the Digital Audio Visual Council (DAVIC).

3.2.1 IETF

Work related to SAS is covered by four different IETF working groups: *Audio/Video Transport, Datagram Congestion Control Protocol, Multiparty Multimedia Session Control*, and *Content Distribution Internetworking.* While the first three working groups belong to the *Transport Area*, the last one is part of the *Application Area.*

3.2.1.1 Audio/Video Transport (AVT)

IETF's *Audio/Video Transport* working group (AVT) is mainly concerned with the specification of real-time transport protocols for audio and video. Work in this group led to the development of the RTP [62] standard and the specification of a series of payload formats for RTP [63]. The standards of this working group are well known and integrated into many commercial and open source video streaming and conferencing applications. Recently, the working group drafted a new version of the RTP RFC [64] which is mainly concerned with changes to rules and algorithms and leaves the packet format on the wire unchanged. It is mainly the timer algorithm that calculates when to send RTCP packets that is changed. Thus, these recent changes have no impact on the SAS architecture.

3.2.1.2 Multiparty Multimedia Session Control (MMUSIC)

The control of multimedia communication in the Internet is addressed by the *Multiparty Multimedia Session Control* (MMUSIC) working group. For the control of audio and video data, the *Real Time Streaming Protocol* (RTSP) [16] has been specified while multimedia sessions are described by the *Session Description Protocol* (SDP) [65]. Recent activities in this group are concerned with the creation of a new RTSP version [66] which is mainly concerned with fixing existing flaws in the standard and the next generation SDP (SDPng) [67] which deals with fundamental changes in SDP. Issues related to overcoming security flaws in RTSP with the aid of firewalls are presented in more detail in references [68] and [69].

3.2.1.3 Datagram Congestion Control Protocol (DCCP)

The relatively new *Datagram Congestion Control Protocol* working group (DCCP) deals with the development and specification of a *Datagram Congestion Control Protocol* [70]. Their goal is to develop a protocol that establishes congestion control for an unreliable packet stream. The way DCCP is designed it must be seen as a framework which determines the general rules of the protocol behaviour and a general message format but allows the usage of different congestion control algorithms which are separately specified as profiles. One of these profiles [71] specifies the usage of *TCP-friendly Rate Control* (TFRC) as a congestion control algorithm in DCCP.

As an alternative the *Stream Control Transmission Control Protocol* (SCTP) [72] could be used. It was originally designed to transport public switched telephony network (PSTN) signalling messages over IP networks, but it is also capable of supporting other applications such as video streaming. SCTP is a reliable transport protocol that operates on top of IP. The latter is the major drawback of SCTP. Making use of SCTP would require kernel extensions, since it is not integrated into most standard operating systems.

3.2.1.4 Content Distribution Internetworking (CDI)

New standards that specify the interoperation of separately administered CDNs are defined by the *Content Distribution Internetworking* (CDI) working group. The goal of this working group is to specify the requirements for CDI rather than the development of new protocols. The group is mainly concerned with the definition of requirements for content distribution internetworking, namely interoperation of request-routing systems, interoperation of distribution systems, and interoperation of accounting systems. An introduction to CDNs, CDI and the specification of a common vocabulary is given in reference [73]. A more detailed overview of the activities of this working group can be found in reference [74].

3.2.2 DVB and DAVIC

The Digital Video Broadcasting (DVB) project was started in 1993 with the goal of developing digital terrestrial TV in Europe. Over the years the work in DVB also covered satellite and cable networks. Since the work in DVB is mainly focused on broadcasting it is not directly applicable to VoD services. Nevertheless, there exist VoD services that make use of the underlying DVB technology. For example VoD services, such as that offered by Kingston [75], make use of IP on top of DSL and use only the existing infrastructure such as cable. To be able to support IP, new routers must be added in these networks. This enhanced infrastructure allows the usage of streaming solutions which are devised for IP-based networks. Applications that make use of the new mechanisms presented in this book can be applied to such networks. Thus DVB can be seen as an underlying technology that offers an infrastructure which is used by the video distribution system.

The Digital Audio-Visual Council was founded in 1994 with the goal of specifying open interfaces for interactive digital audio-visual services. Shortly after its foundation the interest in DAVIC from industry and research was high. With the increasing popularity of the Internet and the standardization of new protocols for streaming [66, 62] the focus shifted. This development led to the situation that DAVIC is no longer active. The goal to achieve open interfaces for audio-visual services could not be reached. Today, available video streaming systems that are used in the Internet are not fully compatible [76]. Some of them support a minimum set of compatibility by using the same control protocol (RTSP), but it is not assured that the client of vendor B can communicate with a server or cache of vendor A.

3.3 CONTENT SCALABILITY–SCALABLE ENCODED VIDEO

In section 2.4.5 content scalability was introduced as one major component of SAS. It is required to adapt the transmission rate of a streaming session to the available bandwidth on the network. In this section an overview of the work on scalable encoded video is given. Figure 3.1 shows the difference between non-scalable and scalable content in the case of video data.

Basically there are four different classes for scalability in the case of video. The first three are spatial scalability temporal scalability and SNR (signal-to-noise ratio) scalability [77] (see Figure 3.2). Spatial scalability offers the functionality to decode images at different spatial resolutions, while temporal scalability allows the adjustment of the frame rate. SNR scalability is achieved, for example, through the layered quantization of the DCT values [78] or other methods like the *embedded zerotree wavelet* (EZW) [79] algorithm. A new, fourth, class of scalability, emerged with the development of the MPEG-4 standard which allows the composition of a video scene from several independent video objects. Since this also makes it possible to decode a subset of objects, *object-based scalability* is introduced as shown in Figure 3.2.

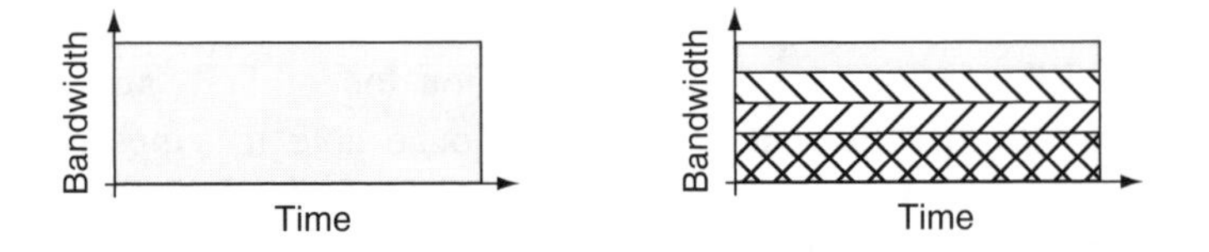

Figure 3.1 Non-scalable vs. scalable content.

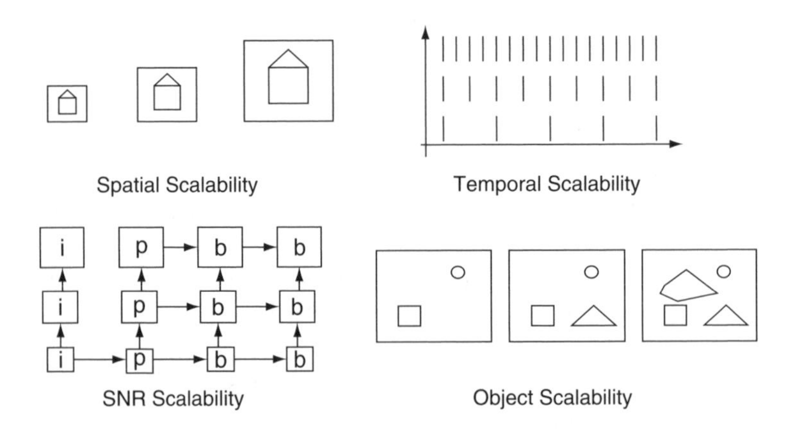

Figure 3.2 Classes of scalability.

In general, as with encoding formats like MPEG-1, only data loss up to a certain limit allows the reconstruction of a frame. This is different if scalable encoding schemes are applied. In this case, part of the video data is sufficient to reconstruct the video signal with the trade-off that the quality of this video signal is reduced. Additionally, the coding efficiency of scalable encoding approaches is reduced, since a higher amount of redundancy information is needed compared to non-scalable encoding formats.

3.3.1 Hierarchically Layer-encoded Video

An encoding scheme that makes use of scalability is layered encoding. With layered encoding the video is split into one *base layer* and one or more *enhancement layers*. The base layer contains fundamental coding information and can be decoded without any additional information. Enhancement layers contain additional information that increase the quality of the reconstructed video signal. In contrast to the base layer, enhancement layers are not independent of other layers. To reconstruct the information included in layer, n all of the information of the lower layers $(0, \ldots, n-1)$ are needed. If the base layer is missing, no video signal can be reconstructed at all. An introduction and overview about layered encoding techniques is given in more detail in [80]. One specific example for layer-encoded video is SPEG [81] which is used by me for the assessment of variations in layer-encoded video. A detailed description on SPEG is given in section 4.3.1.

The concept of scalable coding was first introduced in the MPEG-2 [82] and H.263 [83] standards, which allow a two-layer encoding (base layer plus one enhancement layer). This was extended with the H.263+ [83] standard which allows several layers.

3.3.2 Fine Granularity Scalability

Fine granularity scalability (FGS) is a different kind of hierarchically layer-encoded video. While in traditional hierarchical layer-encoded video all layers are discrete in size, this is different for FGS.

FGS [77] is basically a hierarchical two-layer scheme but allows a variable rate enhancement layer (see Figure 3.3). The rate for the base layer is fixed, similarly to traditional hierarchical layer-encoded video. FGS is one of the 'streaming video profiles' specified in the MPEG-4 standard [84].

With FGS the enhancement bitstream can be truncated into any number of bits within each frame. Thus, the rate can be exactly adapted to the available rate of the transmission channel. The quality of each frame is proportionally increased with each decoded bit. This means, in contrast to a discrete layer approach, that the quality does not increase in steps but in a much finer granular way if the transmission bandwidth increases. In the case of discrete layers, the enhancement layer will only be decoded if all bits of a specific frame arrive at the client. Therefore, only in the case of a bandwidth increase that is of exactly the enhancement layer rate or higher can a quality improvement be perceived.

FGS coding is realized in the way that the base layer is encoded with the MPEG-4 DCT-based standard, while for enhancement layer encoding bitplane DCT coding is used. This method generates variable length codes which build the FGS bitstream. For example, all most significant bits from the DCT coefficient of the enhancement layer of a video block form bitplane 0 and the second most significant bits build bitplane 2, etc. Sequentially arranging these bitplanes allows an arbitrary truncation for each enhancement layer frame when streamed on the network.

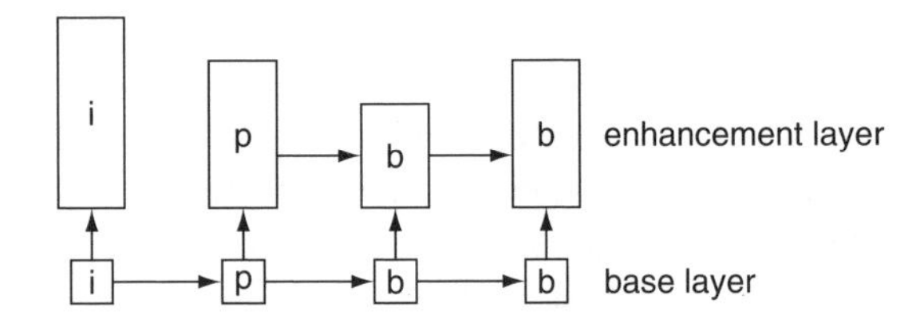

Figure 3.3 Fine granularity scalability.

Recently, FGS has become popular in video streaming applications and professional FGS encoders and decoders are widely available.

The disadvantage of both schemes is the fact that motion vector information is only included in the base layer. Thus, the loss of base-layer information results in the loss of all data for a specific frame because of the lack of motion vector information.

3.3.3 Multiple Description Coding

A new scalable scheme that has gained attention recently is *multiple description coding* (MDC) [85]. MDC codes a video into two or more descriptions and either description can be used to decode baseline quality video. It can also be categorized as a layered encoding scheme with the advantage that the layers (i.e., descriptions) are independent of each other, which is in contrast to hierarchical layer encoding. All layers are of equal importance. Thus data loss of one description does not result in a complete frame loss. The quality of the decoded video increases with the number of descriptions received at the client.

Interestingly, MDC was originally developed as an encoding scheme for audio to be applied in the standard, wired telephone system [86]. Here, the idea was to provide two different routes between two endpoints and to send each description by a separate route. Thus, in the case of a line failure a call, although with diminished quality, would still be possible. With the increasing popularity of video streaming the MDC principle was also applied to video encoding.

It has been shown that MDC is well suited for networks that allow *path diversity* [87]. That means the single descriptions can be sent via separate paths to the receiver, thus preventing congestion on one path from affecting all descriptions. Path diversity in the Internet is hard to achieve, since source routing is not widely supported. Using MDC as a scalable streaming format for video distribution systems in the Internet can be beneficial, but only if path diversity is also supported. Future solutions may achieve path diversity by storing each description on a different server, as proposed by Apostolopoulus *et al.* [87]. Yet, there are some issues that have to be investigated in further detail. Simply using different servers for each description does not necessarily mean that both paths, between the servers and the client, differ much. In addition, the proposed scheme allows only a maximum of two descriptions, which is rather restrictive. Figure 3.4 compares a possible scenario at which layered, FGS and multiple description encoded video would be transmitted over a best-effort link.

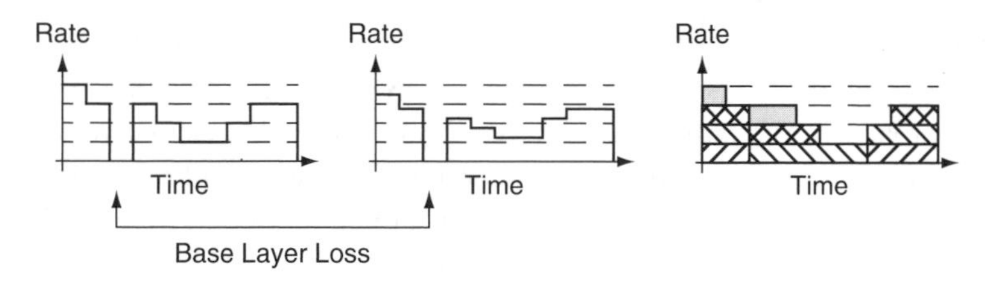

Figure 3.4 Layered, FGS and MDC video transmission.

3.3.4 Comparison of Layered Encoding Approaches

All of the layered approaches have advantages and disadvantages when compared with each other.

It is obvious, that the coding efficiency of all layer-encoding approaches is worse than a non-layered, monolithic format. This means, an MPEG-4 stream which is encoded in a non-layered format has a lower rate for the same quality as a MPEG-4 FGS stream. For equal rates the PSNR of the FGS video can be up to 2 dB lower compared to the non-layered video [84]. Non-layer coding has the highest coding efficiency owing to the fact that the compression procedure for the layered approaches is more complex. With the aid of MDC this phenomena can be explained quite well. In MDC all descriptions have to include motion vector information while in a non-layered approach this information must only be available once. Comparing all three of the above-presented approaches reveals that FGS has the best coding efficiency, followed by discrete, hierarchical layer-encoding, and MDC has the worst coding efficiency. FGS's better coding efficiency is caused by the fact that the bitplane coding of the DCT coefficients in FGS is more efficient than the usual run-length coding. Thus the average PSNR of an FGS video can be up to 2 dB higher [84] compared to a discrete, hierarchically layer-encoded video.

In the case of zero or almost no losses, the layered approaches are more applicable. The precautions that are made to deal with losses in the base layer do not justify the lower coding efficiency introduced with MDC encoding. Lee *et al.* [88] have shown that layered-coding performs better than or equal to MDC if losses are below 2%. The result of video streaming experiments conducted across the USA in 2002 revealed an average loss rate of less than 1% [89]. If, in the future, streaming applications in ad hoc networks become popular, MDC would become an attractive encoding alternative. Owing to the very unstable network topology in ad hoc networks, descriptions could be sent via different paths from the sender to the receiver. At least one description would arrive at the receiver in the case that one path completely fails.

The comparisons made here show that FGS is the most favourable encoding approach in the case of no, or very low, packet losses. There are two reasons for this: (a) it has the highest coding efficiency of all three presented encoding schemes, and (b) it can adapt its streaming rate to every available transmission rate that is greater than the base layer rate. Throughout the book it will be shown that FGS is not always the best solution; it depends on the application scenario which solution is best.

It should also be mentioned that FGS can always be streamed as a discrete, hierarchical layer-encoding scheme. In this case, the sender would only increase the quality if the transmission rate exceeded a certain threshold. Here, the rate would be increased by a significant amount. If the transmission rate falls below a certain threshold, the streaming rate would be decreased in a larger step. In Chapter 4, it will be shown why the application of such a transmission scheme might be beneficial.

3.4 CONGESTION CONTROL–TCP-FRIENDLINESS

Besides a scalable distribution infrastructure, it is very important for an Internet TVoD system to take into account the 'social' rules implied by TCP's cooperative resource management model, i.e., to be adaptive in the face of an (incipient) network congestion. Therefore, the streaming mechanisms of an Internet TVoD system need to incorporate end-to-end congestion control to prevent unfairness against TCP-based traffic and to increase the overall utilization of the network (Figure 3.5).

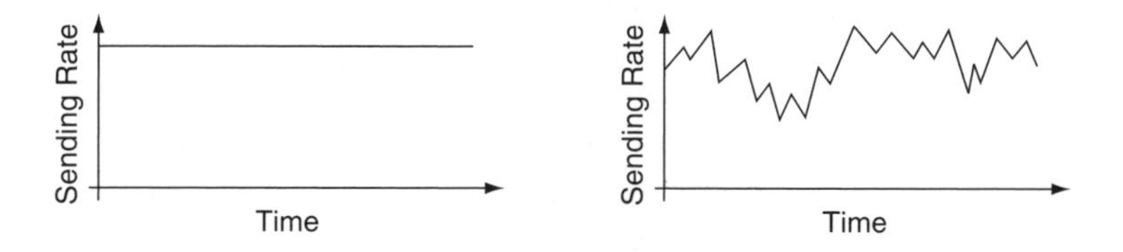

Figure 3.5 Non-congestion-controlled vs. congestion-controlled.

In recent years several protocols for the transport of non-TCP traffic that incorporate TCP-friendly congestion control have been developed. Widmer *et al.* [28] have published an overview of the approaches. To be applicable for streaming these protocols have to meet the following requirements:

- rate oscillations must be kept to a minimum;
- modification to the network infrastructure must be prevented (for example, the protocol stack in the routers may stay as it is).

The existing approaches can mainly be separated into two major categories, window-based and rate-based.

3.4.1 The Window-based Approach

MTCP [90] and pgmcc [91] are two examples of a window-based congestion control approach. In this case a congestion window is maintained similar to the one used in TCP. The window size is increased if no congestion occurs and decreased in the case of congestion. Since this approach is quite similar to the TCP congestion control mechanism, the resulting transmission rate of window-based TCP-friendly protocols fluctuates (sawtooth-like) and, thus, the protocols are not very well suited for the transmission of audio and video streams which require a constant transmission rate. Even the use of layer-encoded video would not solve this problem, since the strongly oscillating transmission rate would lead to a high number of layer changes, which impair the perceived quality at the client (see Chapter 4).

3.4.2 The Rate-based Approach

With rate-based congestion control the transmission rate is determined according to a network feedback mechanism indicating congestion. Rate-based congestion control can be subdivided into the AIMD (additive-increase/multiplicative-decrease) and the model-based congestion control schemes. The AIMD scheme (e.g., RAP [46]) behaves similarly to TCP's congestion control and, therefore, the resulting transmission rate oscillates. This effect leads to the same problems as described for window-based congestion control. In the model-based approach a TCP model is used to determine the TCP-friendly transmission rate. TFRC [47] is a TCP-friendly protocol which makes use of such an approach. The model is based on the TCP throughput equation as shown in (1).

$$\text{Sendrate} \approx \min\left(\frac{W_{\max}}{RTT}, \frac{1}{RTT\sqrt{4p/3} + B_{\min}(1,\ 3\sqrt{3p/4})p(1+32p^2)}\right) \quad (1)$$

where $B =$ timeout, $p =$ loss rate and $W_{\max} =$ maximum congestion window size.

The advantage of the model-based approach is the smoothness of the resulting transmission rate. This is caused by the fact that the sending rate is adapted to the long-term TCP throughput.

From experimental observations ([92] and Chapter 7) and the results presented by Widmer *et al.* [28], TFRC is very promising as a TCP-friendly protocol for streaming [47]. It is a rate-based congestion control protocol with TCP-friendliness over longer timescales. The main advantage in combination with A/V streaming is that the rate is smooth in the steady-state case and, therefore, applications that rely on a constant sending rate are supported. In addition, the protocol is end-to-end based which does not require any modifications of the network infrastructure. TFRC has the advantage over TCP that its transmission rate is not limited by any window size. TFRC is the most prominent of all TCP-friendly congestion control approaches and has recently been considered to become a standard [71] in the IETF. TFMCC [93] is an extended version of TFRC that supports single-rate multicast.

Besides the approaches that were evaluated in reference [28] there is also related work done by Bansal and Balakrishnan [94] called binomial congestion control which is a nonlinear generalization of AIMD. In the case of a packet loss the window is reduced from W to $W - bW^l$ and an RTT without loss leads to $W + a/w^k$. For $l = 0$ and $k = 1$ the algorithm is equivalent to TCP's AIMD.

Additional work with the main aspect to investigate TCP-compatibility by simulation and analysis is presented by Bansal *et al.* [95]. The results of this work show that all the investigated algorithms (RAP, TEAR, TFRC and binomial congesting control) can be deployed in the Internet, since they behave in a TCP-compatible way. In addition, the investigation revealed drawbacks of some of the algorithms and led, for the case of TFRC, to a modification that improves TFRC's behaviour in the case of abrupt bandwidth reduction.

3.5 ADAPTIVE STREAMING–STREAMING LAYER-ENCODED VIDEO WITHOUT CACHES

The work that has been performed on streaming of layer-encoded video can be subdivided into two categories. In the first, layer-encoded video is used in combination with congestion-controlled streaming. (The difference between non-adaptive and congestion-control based adaptive streaming is shown in Figure 3.6). Rejaie *et al.* [96] make use of RAP [46] to stream layer-encoded video in congestion-controlled manner. In combination with a buffer that is located at the client they address the problem of absorbing short-term fluctuations in bandwidth that are introduced by the congestion-control algorithm (sawtooth shape). Thus, buffering a few seconds of a stream can lead to smoother playout because of reduced layer changes. Quite similar to this work is that presented by Nelakuditi *et al.* [97] that in addition to [96] uses quality metrics and presents new algorithms that maximize those metrics. A similar investigation has been performed by Feamster *et al.* [98], with

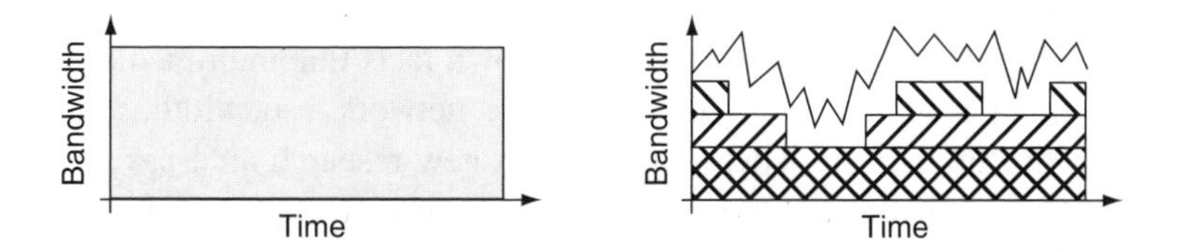

Figure 3.6 Non-adaptive vs. adaptive streaming.

the difference that the binomial congestion-control algorithm [94] is used. The authors show that their congestion-control algorithm performs better in terms of buffer usage and average target bitrate in comparison to the one presented by Rejaie. In reference [99] the available bandwidth in the network is modelled as a stochastic process and in comparison to the work of Rejaie and Nelakuditi only layer-encoded video consisting of two layers is regarded and the client's buffer is unlimited. Another proposal for adaptive streaming is given by Tan and Zakhor [48]. In addition to their TCP-friendly transport protocol they also present a new scalable encoding scheme that overcomes the shortcomings of hierarchically layer-encoded video. This encoding scheme allows the decoding of upper-layer data despite the absence of base-layer information (similar to the MDS approach described in section 3.3.3).

In addition to the work presented above, investigations on layer-encoded streaming in the context of multicast distribution have been performed [100, 101].

One of the first applications of scalable content has been receiver-driven layered multicast [100] which allows many heterogeneous receivers to take part in a multicast streaming session. This application was designed more for the streaming of live events and, therefore, interactivity is not supported. To enable a higher level of interactivity, which is needed to support TVoD, multicast can only be applied in a limited scope. Thus, caches gain even more importance in a distribution system in the drive to reduce the server's load. In addition, making use of scalable content such as layer-encoded video is also a valid means in video distribution systems that support TVoD and have to adapt to current network conditions and heterogeneous clients.

3.6 SYSTEM SCALABILITY–CACHES

The work on proxy caching for web objects has received much attention in recent years. (Figure 3.7 shows the difference in a distribution architecture with and without caches.) This work has mainly been concerned with the caching of HTML pages and still images. Further studies have revealed that there is a fundamental difference between the caching of conventional web objects and the caching of multimedia objects such as audio and video streams [26]. This results from the well-known facts that multimedia objects require more storage space, consume more network bandwidth, and have inelastic traffic characteristics. Therefore, a new research area has emerged in recent years that is concerned with the caching of multimedia objects in the Internet. An overview of the most relevant work in this area is presented in the following.

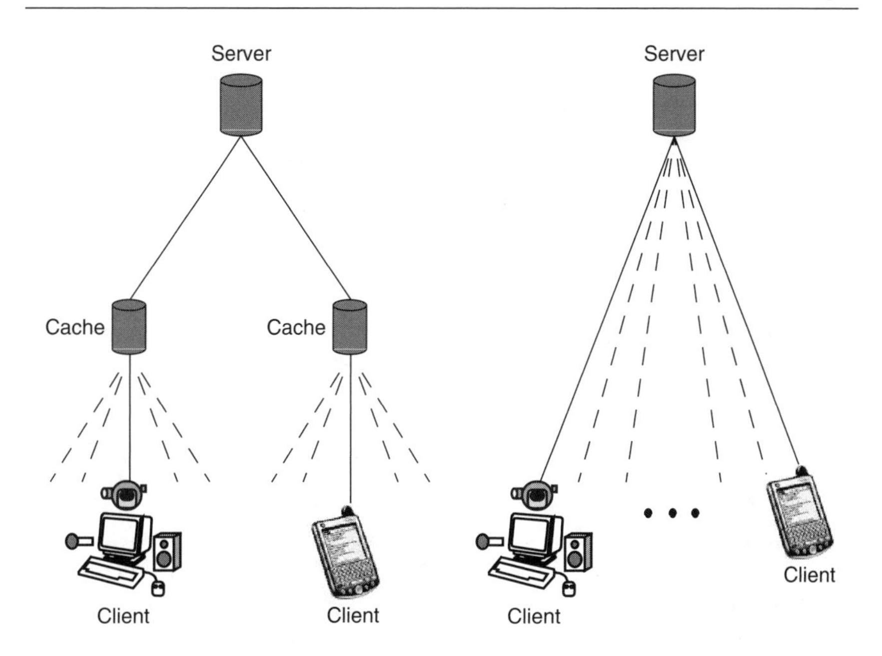

Figure 3.7 Distribution system with and without caches.

This section reflects the requirements on caching and its related mechanisms in the scope of SAS. First of all, a general overview of the work on partial caching of video objects is given. Adaptive streaming in combination with write-through caching (see section 2.4.4) can lead to the situation that video objects are cached only partially. Also several proposals that are focused on the caching of layer-encoded video are presented. Furthermore, mechanisms that achieve a reliable transport into caches are presented, since SAS requires a reliable transmission protocol (see section 2.4.4).

Clustering of caches is an approach to increasing the scalability of caches and, thus, increasing the overall scalability of the video distribution system. The work presented in this book does not focus on cache clusters, but existing related work shows how clusters can be used to increase system scalability.

For a comprehensive overview on complete object caching, the interested reader is referred to reference [31].

3.6.1 Partial Caching of Video Objects

The existing approaches for partial caching can basically be divided into two subcategories, since there are two dimensions for partial caching. The first dimension is time and the second is bandwidth. As shown in Figure 3.8,

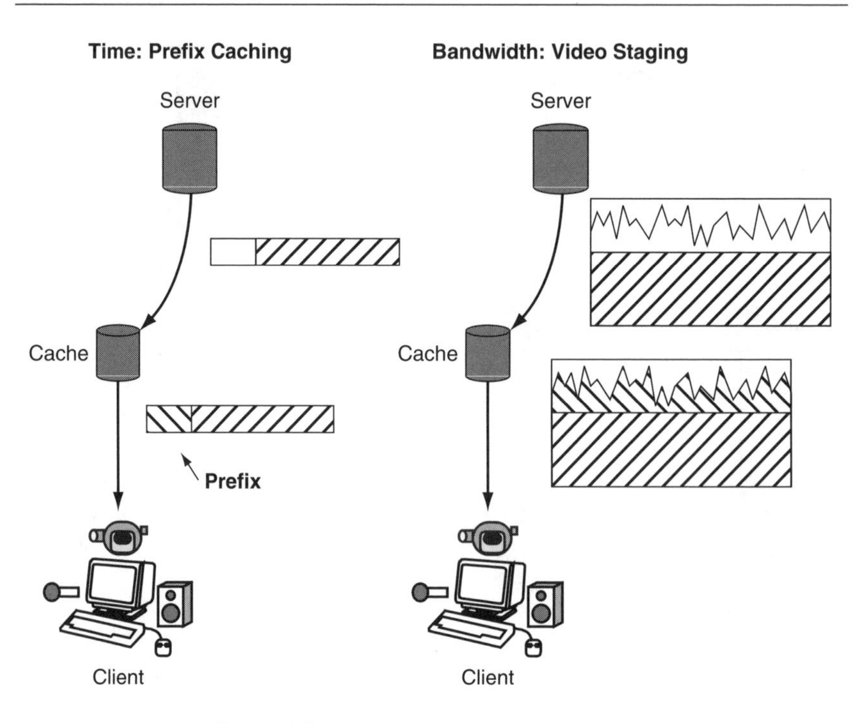

Figure 3.8 Prefix caching and video staging.

only a temporal part of the video is stored, which is in contrast to a second approach where only a fraction of the bandwidth for the whole length of the video is cached.

3.6.2 Time-based Partial Caching

Among the first who performed investigations in this area were Tewari *et al.* [102]. Their *resource-based caching* scheme is intended for non-layer-encoded video objects which can be cached either partially or completely. The motivation to cache fragments of an object is to optimize the cache space usage based on the popularity and the inter-arrival times of requests for an object. Another scheme for partially caching monolithic video objects is the *prefix* caching scheme presented by Sen *et al.* [59]. It is the goal of prefix caching to reduce start-up latency and to smooth rate variations on the link between the server and the cache. The first goal is achieved by storing the beginning (prefix) of a video object at the cache. Immediately with the playout of the prefix the remaining part of the video is streamed from the server to the cache. Depending on the length of the cached prefix the suffix can be buffered at the cache and, thus, smoothing can be applied to achieve a constant transmission rate between server and cache. The length of the prefix that is stored on the cache increases with the object's popularity. *Segment-based* [103] caching is an extension of *prefix* caching. In this approach, the video object is split into segments whose size increases exponentially with the distance from the start of the video. Thus, it can be assured that the prefixes of many objects are stored on the cache while only the most popular objects are cached entirely. Balafoutis *et al.* [104] investigate the impact of the replacement granularity on the overall performance of a cache by varying the chunk size, i.e., the smallest unit of a video that can be stored on a cache. Their simulations show that a smaller chunk size leads to a better performance of the cache. Chan and Tobagi [105] present an extension of prefix caching that is designed for a hierarchical distribution infrastructure. Their investigation assumes that part of the length of each video is temporarily stored in each cache of a hierarchical distribution system. In their *pre-storing* scheme, a request for video m_i that arrives at a cache at level one is answered by first sending the prefix of length v_i that is stored locally. The level-one cache issues the request for the remaining portion of the movie to the level-two cache, forwards this data to the client after playing out its own prefix, and caches the portion from the level-two cache temporarily in a sliding window from which all other clients are served that have requested the same video. While this approach is theoretically highly efficient, it requires that several caches cooperate for a single streaming session.

While the *prefix* schemes and its variations presented above always store contiguous parts of the video on the cache, Miao and Ortega [106] present a method called *selective* caching that stores arbitrary segments in addition to the prefix on the cache. In the case of pre-encoded video, choosing the right packets can increase the robustness of the entire video stream against congestion on the link from the server.

One further extension of *prefix* caching is *MCache* [107] an approach that combines multicast transmission and *prefix* caching. It can also be seen as a modification of the *patching* technique [40] where the patch is streamed from the prefix that is stored at the cache instead of the original version where both patch and multicast stream are sent from the server. An extension of *MCache* is presented by Wang *et al.* [108] who modify the caching mechanisms in a way that allows *MCache* to work also in backbones that do not support multicast transmissions. Request merging by a window-based caching approach is proposed in references [109] and [105]. That is, requests for the same stream arriving relatively close to each other are merged by caching a sliding window of data belonging to a video on the cache. Another window-based approach is the one presented by Rexford *et al.* [110]. In their case, the cache is used as a buffer (smoothing window) to allow smoothed transmission of variable bitrate (VBR) video in the access network between cache and client. Two further approaches that combine multicast transmission with caching are the scheme presented by Verscheure *et al.* [111] and *gleaning* [4]. The goal of these two approaches is to reduce the number of streams that are sent to the client in parallel and to reduce the buffer requirements at the client. Thus, the buffering and sequentialization of the different streams (patch and multicast stream) are performed at the cache and only one stream is forwarded to the client. This is very efficient in environments where access bandwidth is scarce and the client buffer is very limited, as would be the case in a wireless scenario. An analytical model to investigate optimal caching strategies for video objects in combination with segmented multicast delivery is presented by Eager *et al.* [112]. Their focus is on how the heterogeneity of client populations and cache capabilities influence the overall system performance. Results of this analysis show that it is almost always beneficial to store initial segments of many files than to store all segments of fewer files. This is caused by the applied multicast delivery technique where initial segments are repeatedly transmitted very frequently, and caching those segments reduces the load at the server. Somewhat surprisingly, even in the case of systems with heterogeneous features, it is efficient to store the same data set at all caches. A refinement of this work is presented in reference [113]. In this case, a new, more efficient multicast delivery protocol [41] is used in combination with several delivery scenarios. The overall results of this analysis state, in contrast to the previous work, that it is efficient to cache complete objects instead of segments. In cases of high object request rate, multicast delivery without cache is more cost-efficient.

It must be mentioned that not all caching approaches that make use of some sort of multicast delivery mechanisms are designed for the use of streaming and distributing content-scalable video formats such as layer-encoded video owing to the lack of investigations into how to apply those schemes to content-scalable video.

3.6.3 Bandwidth-based Partial Caching

Another approach of partial caching is the one proposed by Zhang *et al.* [114] with their *video staging* system. In their case, a monolithic VBR video is split into two parts. As shown in Figure 3.8, the part of the video stream that exceeds a certain threshold rate is cached at the proxy while the lower-rate part is stored at the server. In the case of a client request the lower-rate part is streamed from the server and the higher-rate part from the cache, thus the backbone bandwidth requirement is reduced and the transmission rate is constant. In comparison with the approaches presented above, which can be seen as temporal partial caching, *staging* can be seen as an approach for the bandwidth dimension in partial caching. The staging approach is somewhat limited in terms of granularity. Should the bandwidth on the link between server and cache be insufficient to transmit the lower-rate part of the video, the stream cannot be offered to the client.

It is quite obvious that layer-encoded video is well suited for this approach of partial caching. Nevertheless, the work on the caching of layer-encoded video has received only little attention compared to the work on time-based partial caching. Rejaie *et al.* [115] were the first to present an approach for the caching of layer-encoded video. The video is streamed in a congestion-controlled manner (using the RAP protocol [46]) from the server through the cache into the client. Missing segments on the cache, caused by losses and rate adaptation, are prefetched in a demand-driven fashion to improve the quality of the cached video. A cache replacement algorithm is presented that works on a fine-grained level which allows the dropping of single segments of a layer. Simulations reveal that the quality of a cached video is directly related to its popularity. Similar investigations have been performed by Paknikar *et al.* [116], with the difference that only complete layers can be dropped. In addition, their approach consists of a cluster of caches which is managed by a broker and is intended for a high-speed local area network.

An analytical investigation of this topic was performed by Kangasharju *et al.* [117]. Their main goal was to gain better insights into the effects cache space and link bandwidth have on the cache performance. In contrast to reference [115], only complete layers can be stored or removed from the cache, in order to keep the problem mathematically tractable. Congestion control on the link between the server and the cache and the cache and the client is not assumed. For the streaming from the server a certain rate is allocated and the transmission of the requested stream starts if this rate is available and it is assumed that this rate is available for the remainder of the streaming session. Thus, this model can only be applied in an environment that gives bandwidth guarantees. In order to make it applicable for the best-effort service, the model of the bottleneck bandwidth has to be modified.

A prototype implementation of an adaptive multimedia cache is presented in reference [118]. In this case the *Squid* web proxy cache [119] is modified in order to perform this kind of caching and perform initial experiments on their prototype. These preliminary results tend to confirm the simulative results from reference [115]. Further experiments are necessary to examine performance and behavior of this prototype cache in more detail.

An analytical and simulative investigation of FGS video caching is presented in reference [120]. In this work, adaptation to the available transmission bandwidth is not considered. However, heterogeneous clients are supported by this architecture. This support is achieved by grouping clients in different classes and assigning a transmission bandwidth to this class. In the case of a group of three clients with a receiving capability of 640 kBit/s, 700 kBit/s, and 720 kBit/s, these clients would be served with a 640 kBit/s stream. Based on the available backbone bandwidth and the available space on the cache, a video can be partially cached both time-based and bandwidth-based. The time-based caching is performed similarly to reference [59]. For the bandwidth-based caching the cache can act like a filter if the video cannot be cached in the bandwidth as it is streamed to the client. That means, the video is cached in a smaller, but constant, bandwidth than it is forwarded to the client in. This is in contrast to SAS where the video is not necessarily streamed with a constant transmission rate and, thus, variations in the quality of the cached video can occur.

3.6.4 Disadvantages of Partial Video Caching

The disadvantage of not caching video objects entirely is the reduced fault tolerance. The failure of the server or the link from the server to the client or the cache would lead to a complete failure of the streaming service. This is in contrast to approaches that store a complete video object with the drawback that more storage space is needed at the caches. There clearly exists a trade-off between fault tolerance on the one hand and the efficiency of a cache (in terms of storage space consumption) on the other. An additional drawback of partial caching in the temporal dimension is the limited support for interactivity such as VCR functionality. For example, in the case of *prefix* caching a jump to a position further ahead in the video leads to increased latency, since the entire data has to be streamed from the server.

Approaches that make use of layer-encoded video can be seen as an exception if the single layers are always cached completely. In this case, a server or link failure does not necessarily lead to a service failure but can lead to a quality decrease of the service, since not all layers might be stored on the cache.

It should be mentioned that all of the presented approaches make assumptions on the reliability of the link between the server and the cache. The most popular prefixes used in *prefix* caching and its variations, for example, can be actively pushed from the server to the cache via a reliable transport protocol such as TCP. The cache can also obtain the prefix by caching the first segments of a stream that is initially requested by a client if the stream is forwarded through the cache. In the latter case the transport would be unreliable, and it cannot be assured that the prefix that is identical to the one stored on the server is cached. For reasons of simplicity, all approaches assume that losses between the server and the cache do not occur and, therefore, do not provide any loss-recovery mechanisms. In today's best-effort Internet, loss-recovery mechanisms have to be applied in order to make the proposed approaches realistic. In the following section mechanisms to achieve reliable transport of video data into caches are presented.

3.7 RELIABLE TRANSPORT INTO CACHES

One of the major goals in an environment for A/V-caching should be to obtain a cached version of the content in the cache that is similar to the original content to avoid error propagation towards the client. With the use of standard RTP based on UDP, information that gets lost during transmission (as shown in Figure 3.9) is also lost to the caches. The problem is that these errors would be transmitted with every stream that is forwarded from the cache server to a client. In any case that should be avoided since it has to be regarded as a degradation of the service quality.

The work on reliable multicast led to the development of a series of reliable multicast protocols on top of UDP. Using reliable multicast in a video distribution infrastructure bears the advantage that simultaneous transport of video objects into caches can be executed more efficiently than in the case of uni-cast. Some examples for reliable multicast protocols are SRM (Scalable Reliable Multicast, [121]), TRM (Transport Protocol for Reliable Multicast, [122]), RMTP (Reliable Multicast Transport Protocol, [123]) and LRMP (Light-weight Reliable Multicast Protocol as an extension to RTP, [124]). TRM and LRMP make similar assumptions about loss detection and repair requests as SRM, so SRM can be discussed as an example for all three protocols. RMTP provides sequenced lossless delivery of bulk data (e.g. Multicast FTP), without regard to any real-time delivery restrictions. It is not applicable for streaming applications, because the retransmission of the missing data is done immediately after the loss detection.

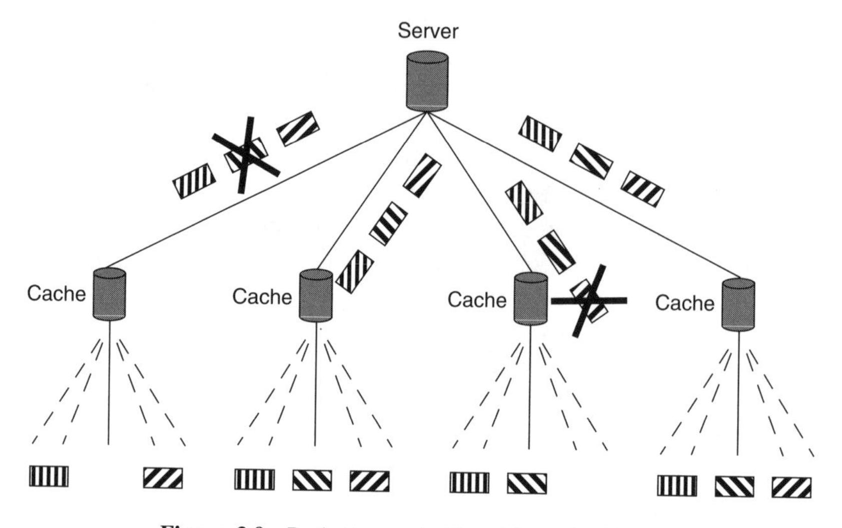

Figure 3.9 Packet losses in IP multicast distribution.

SRM is a reliable multicast framework for light-weight sessions and application level framing. Its main objective is to create a reliable multicast framework for various applications with similar requirements on the underlying protocol. Each member of a multicast group is responsible for loss detection and repair requests. The repair requests are multicast after waiting a random length of time, in order to suppress requests from other members sharing that loss. As it is possible that the last packet of a session is dropped, every member multicasts a periodic, low-rate, session message including the highest sequence number. It must be mentioned that SRM needs a specific distribution infrastructure which is not widely available in the Internet at the moment.

A third class of reliable multicast protocols are the ones which include FEC (forward error correction) as a technique to achieve reliability [125]. Reliable multicast achieved through FEC is also applicable for streaming systems, since usually no retransmissions are necessary during the multicast transmission. The major drawback of this approach is that error-correction information appropriate for the client with the worst connection must be included in each multicast packet. This leads to a higher use of bandwidth, thus leading to a reduced connection quality for the clients. In addition, a completely new protocol must be built in the case of layered FEC, since this model is not compatible with already existing protocols.

All of the aforementioned approaches either need a specific infrastructure and/or additional functionality in the clients and are not designed for distribution systems that encounter caches as part of the infrastructure.

SR-RTP [126] is a somewhat different loss recovery approach designed for unicast delivery of MPEG-4 video. Retransmissions of lost data are performed on the basis of a certain priority. The priority depends on the content of the lost packet. If the missing data belongs to a reference frame (I-frame) it is requested with a higher priority than data that belongs to dependent frames (P- and B-frames). Based on the priority it might occur that some of the missing packets are never requested for retransmission. Thus, SR-RTP is well suited to improve the quality of a streamed video if only client and server are involved but cannot be used in combination with caches. With SR-RTP it is not guaranteed that an identical copy of the original video object is created on the cache.

An additional approach to achieve reliable transport of video data into caches is LC-RTP [127]. It is not only applicable to multicast transmission but can also be used in the case of unicast. On of the major design goals for LC-RTP was the realization of lossless transport of video data into caches while the stream is also concurrently received by the client. Since LC-RTP is a standard compliant extension of RTP, non LC-RTP capable clients can also receive an LC-RTP stream. The SAS architecture builds on LC-RTP to achieve a reliable transport of video data into caches. A detailed description of LC-RTP is given in Appendix A and an extension to support layer-encoded video is presented in section 8.4.1.

3.8 CACHE CLUSTERS

The goal of clustering (an example for a cache cluster architecture is shown in Figure 3.10) is to distribute the load introduced by client requests on to several single servers. The clusters are usually represented as one single entity. Thus, the client is not aware of the cluster. Next to the advantage of load balancing, video server clusters have the advantage of an increased fault tolerance. If one of the servers crashes, existing sessions can be redirected to other servers and, thus, allow a continuation of the streaming process. To allow load-balancing or an increased fault tolerance either a shared storage for the servers of a cluster must exist or data must be replicated between the single servers.

Cache clusters do not necessarily offer the functionality mentioned above. They are mainly used for cooperative caching. Accordingly, data between

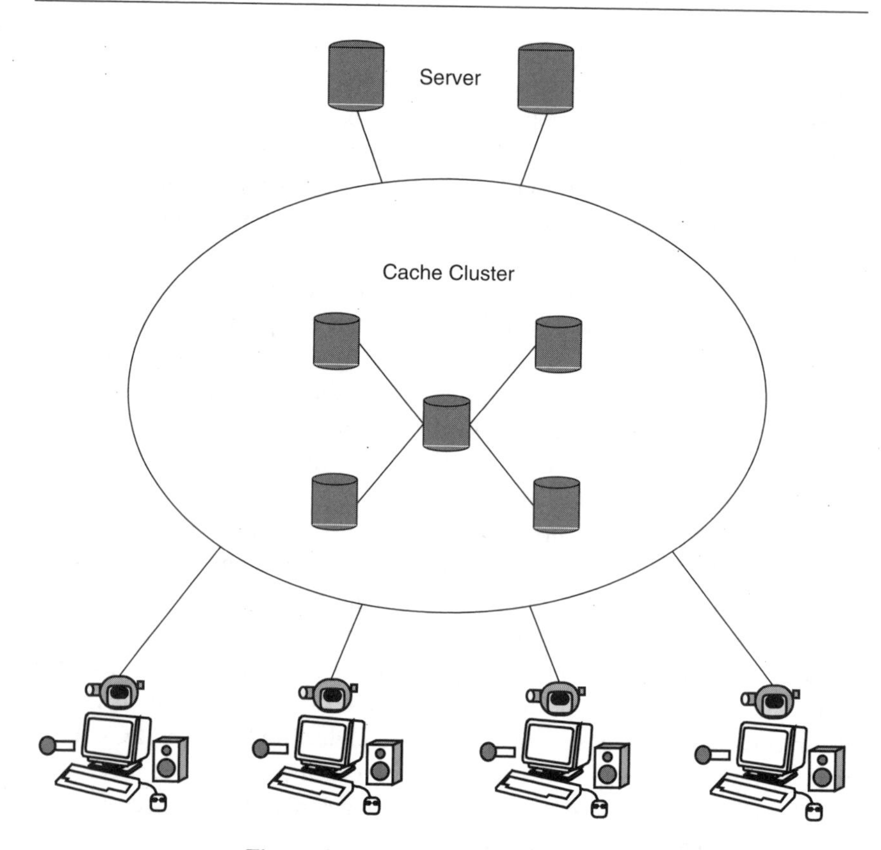

Figure 3.10 Cache cluster architecture.

caches is not necessarily replicated or they do not share the storage medium. Still, a certain level of load-balancing can be achieved depending on how single video objects are distributed for storage on the single caches.

In section 3.6, an approach for cooperative caching was specified for layer-encoded video. Another approach that allows the storage of segments (which belongs to the group of caching methods presented in section 3.6) but not layer-encoded video is presented in the *MiddleMan* architecture [128]. As in [116] all cooperative proxies must be located in a single, well-provisioned LAN and a central *coordinator* is needed to determine where to cache a segment or from which of the local caches to fetch a single segment. A simulative analysis of the *MiddleMan* approach is performed that shows the applicability and benefits of such a cooperative caching method for video objects.

A distributed caching architecture that does not make use of a central coordinator or broker is the one presented by Chae *et al.* [129]. This approach is mainly intended for LANs and MANs since segments of a video object can be stored in a distributed fashion over several caches of the cluster. The segmentation of the video object and the distribution of the segments are based on their *Silo* algorithm. To minimize start-up latency the first segment is stored on all caches of the cluster while the probability of storing further segments on a cache decreases. Thus, the higher the segment number the fewer the copies of the segment that exist in the cluster. It is guaranteed that at least one copy of each segment exists in the cluster. The minimized start-up latency is in trade-off with an increased storage requirement, e.g., in comparison to [128]. *Rainbow* and *Caching Token* are the cache replacement strategies that are used in the cache cluster. Analytical and simulation results show that the cache hit rate can be increased by a factor of up to eight in comparison to traditional web caching systems.

Finally, the approach presented by Hofmann *et al.* [109] is not limited to usage in LANs and MANs but also allows caches to be distributed in the Internet. Also here the video object is divided into smaller segments and employs a window-based caching approach (see section 3.6) to merge requests. In addition, requests for a certain object can be redirected to other caches. Scalable state distribution between the caches is achieved by the *Expanding Ring Advertisement* (ERA), a multicast approach in which advertisements are sent with dynamic TTL values, achieving that caches with a further distance from the sender receive updates less frequently.

Castro *et al.* [130] combine a peer-to-peer overlay network with an application level multicast system, called *SplitStream*, which can be used for video distribution. In their approach a video is split into stripes which are distributed via separate multicast trees. The goal of SplitStream is to create a forest of separate multicast trees in such a way that a node is only an internal node for one multicast tree and a leaf node in all other cases. This mechanism distributes the load equally over all nodes of the distribution system. This approach is well suited for the distribution of layer-encoded video, since each layer can be distributed via a separate multicast tree. Preliminary results of a performance analysis are promising, but further investigations are necessary to show the applicability of this approach for video distribution.

4

Quality Variations in Layer-encoded Video

In the area of video streaming layer-encoded video is an elegant way to overcome the inelastic characteristics of traditional video encoding formats such as MPEG-1 or H.261. Layer-encoded video is particularly useful in today's Internet where a lack of quality of service (QoS) mechanisms can make an adaptation to existing network conditions necessary. In addition, it has the capability to support a wide variety of clients while only a single file[†] has to be stored at a video server for each video object. The drawback of adaptive transmissions is the introduction of variations in the number of transmitted layers during a streaming session. These variations affect the end-user's perceived quality and thus the acceptance of a service that is based on such technology.

Recent work that has focused on reducing those layer variations, by employing either intelligent buffering techniques at the client [97, 99, 96] or proxy caches [115, 118] in the distribution network, made various assumptions about the perceived quality of videos with time-varying number of layers. An extensive literature research was performed to investigate whether these assumptions had been verified by subjective assessment. But existing work that exactly meets the abovementioned conditions could not be found.

Based on this lack of in-depth analysis about quality metrics for variations in layer-encoded videos, an empirical experiment based on subjective assessment was conducted. Here the goal was to obtain results that could be used in classifying the perceived quality of such videos.

[†] In contrast to the dynamic stream switching [132] approach where for each quality level one specific video file is required.

4.1 WHAT IS THE RELATION BETWEEN OBJECTIVE AND SUBJECTIVE QUALITY?

The goal of the work presented in this chapter is to investigate whether general assumptions made about the quality metrics of variations in layer-encoded videos can be verified by subjective assessment. The following example is used to explain the intention of this investigation in more detail. A layer-encoded video that is transmitted adaptively[†] to the client might have layer variations as shown in Figure 4.1. Several quality metrics that allow the determination of the video's quality are presented in section 4.2.1. At first, the basics of these quality metrics are discussed. The most straightforward quality metric would be the total sum of all received segments (see Figure 4.1). However, common assumptions on the quality of a layer-encoded video are that the quality is influenced not only by the total sum of received segments but also by the frequency of layer variations and the amplitude of those variations [97, 115, 133]. As shown in Figure 4.1, the amplitude specifies the height of a layer variation while the frequency determines the number of layer variations.

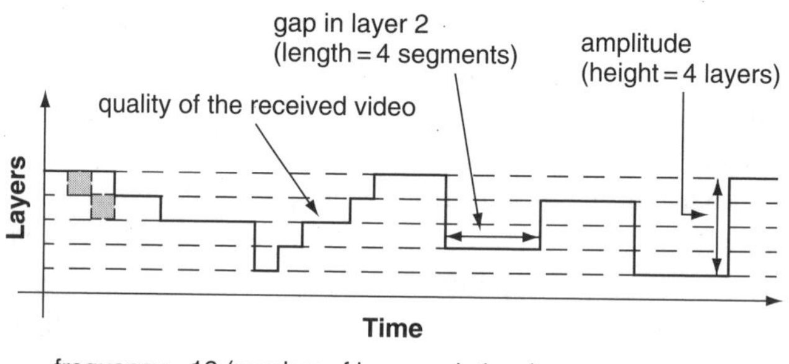

Figure 4.1 Quality of a layer-encoded video at the client.

All known quality metrics are based on these assumptions. Verifying all possible scenarios that are covered by those assumptions with an experiment based on subjective assessment is hard to achieve. Therefore, the focus in this work was set on basic scenarios that have the potential to answer the most fundamental questions, for example, are the sequences on the left in

[†] Adaptively in this case means that the number of layers transmitted to the client is based on some feedback from the network or the client, e.g., congestion-control information.

Figure 4.2 ((a1) and (b1)) more annoying than sequences on the right ((a2) and (b2)) for an viewer who watches a corresponding video sequence. In this example, the first scenario ((a1) and (a2)) is focused on the influence of the amplitude and the second ((b1) and (b2)) on the frequency of layer variations.

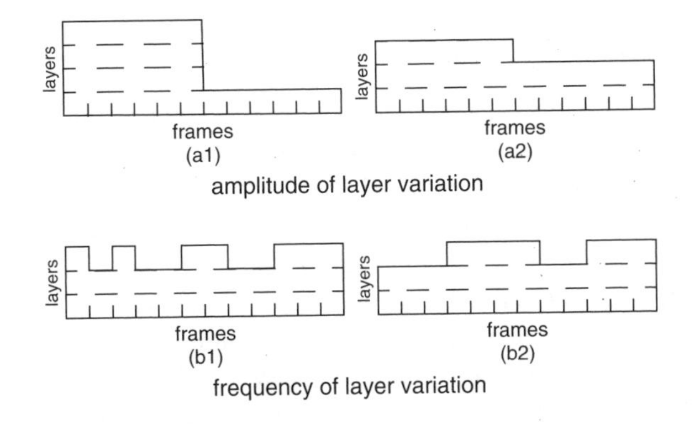

Figure 4.2 Quality criteria [97].

Since the goal of this investigation was to answer some fundamental questions about the influence of layer variations on the perceived quality, the decision was made to perform the subjective assessment on discrete hierarchical layer-encoded video. This decision is based on the fact that this encoding type has similarities to the other two types (FGS and MDC) presented in Chapter 3. Discrete hierarchically layer-encoded video shares the hierarchical aspect with FGS and the discrete aspect with MDC. Thus, results obtained from the subjective assessment presented in this chapter should also be qualitatively applicable for FGS and MDC. Finally, further investigations have to show if the metric presented in section 4.6 can be applied for both encoding types.

This chapter is structured as follows. Section 4.2 reviews previous work on retransmission scheduling for layer-encoded video and subjective assessment of video quality. The test environment and the subjective test method used for the experiment are described and discussed in section 4.3. The details of the experimental setup are given in section 4.4 and in section 4.5 the results of the experiment are presented and discussed. Section 4.8 summarizes the major conclusions that can be drawn from the experiment. A new objective quality metric, called the spectrum, is presented in section 4.6. It is shown that this new quality metric is more appropriate than the peak signal-to-noise ratio (PSNR) metric.

4.2 QUALITY METRICS FOR VIDEO

Given that this work is influenced by the two research areas, quality metrics for layer-encoded video and objective video quality assessment, they are considered separately.

4.2.1 Existing Work on Quality Metrics for Layer-encoded Video

During the investigation of favourable retransmission scheduling algorithms designed to improve the quality of layer-encoded video stored in a cache, it became clear that in related work specific to this aspect the quality metrics for layer-encoded videos are based on somewhat speculative assumptions only, and none of these assumptions are based on a subjective assessment.

Nelakuditi *et al.* [97] state that a good metric should capture the amount of detail per frame as well as its uniformity across frames; that is, comparing the sequences of layers in a video shown in Figure 4.2 the quality of (a2) would be better than that of (a1) which is also valid for (b2) and (b1), according to their assumption. Their quality metric is based on the principle of giving a higher weight to lower layers and to longer runs of continuous frames in a layer. The metric presented by Rejaie *et al.* [115] is almost identical to the one advocated in reference [97]. *Completeness* and *continuity* are the two parameters that are incorporated in this quality metric. *Completeness* of a layer is defined as the ratio of the layer size transmitted to its original (complete) size; for example, the ratio of layer 2 in sequence (a2) in Figure 4.2 would be 1 while the ratio for layer 3 would be 0.5. *Continuity* is the metric that covers the gaps in a layer. It is defined as the average number of segments between two consecutive layer breaks (i.e., gaps). In contrast to the other metrics presented here, this metric is a per-layer metric.

4.2.2 Objective Video Quality Assessment

There has been a substantial amount of research on methods for subjective assessment of video quality (e.g., [134, 135]), which contributed to form an ITU Recommendation [136]. This standard has been used as a basis for subjective assessment of encoders for digital video formats, in particular for MPEG-2 [137, 135] and MPEG-4 [138] but also for other standards such as H.263+ [139]. The focus of interest for all these subjective assessment experiments was the quality of different coding and compression mechanisms. The work presented here, in contrast, is concerned with the quality degradation caused by variations in layer-encoded video. Reference [140] presents the results of an empirical evaluation of four hierarchical video encoding schemes. The focus of their investigation is on the comparison between the different layered encoding schemes and not on the human perception of layer variations.

In reference [141], a subjective quality assessment was carried out in which the influence of the frame rate on the perceived quality is investigated. Elasticity in the stream was achieved by frame rate variation and not by the application of a layer-encoded video format.

The effects of bit errors on the quality of MPEG-4 video were explored in reference [142] by subjective viewing measurements, but effects caused by layer variations were not examined.

Chen presents an investigation, which is based on a subjective assessment, into an IP-based video conference system [143]. The focus in this work is mainly auditorium parameters such as display size and viewing angle. A layer-encoded video format is not used in this investigation.

Lavington *et al.* [144] used an H.263+ two-layer video format in their trial. This is probably closest to the work presented here, although they were rather more interested in the quality assessment of longer sequences (e.g., 25 min). As opposed to using identical pregenerated sequences that were presented to the test candidates, videos were streamed via an IP network to the clients and the quality was influenced in a fairly uncontrolled way by competing data originating from a traffic generator. The very specific goal was to examine whether reserving some bandwidth for either the base or the enhancement layer improved the perceived quality of the video, while the investigation presented here is rather more concerned with the influence of variations in layer-encoded videos and tries to verify some of the basic assumption made about the perceived quality in a subjective assessment experiment. Furthermore, the experiment as described in the following is conducted in a controlled environment in order to achieve statistically significant results.

4.3 TEST ENVIRONMENT

In this section, at first the layer-encoded video format used for the experiment is presented and afterwards the generation of the test sequences is described, the decision to use stimulus-comparison as the assessment method is explained, and finally the test application is presented.

4.3.1 Layer-encoded Video Format–SPEG

SPEG (scalable MPEG) [81] is a simple modification to MPEG-1 which introduces scalability in the transmission rate of a video stream. In addition to the possibility of dropping complete frames (temporal scalability), which is already supported by MPEG-1 video, SNR scalability is introduced through layered quantization of the discrete cosine transform (DCT) data [81]. The extension to MPEG-1 was made for two reasons. First, at the point in time the investigation was performed there were no freely available implementations of layered extensions for existing video standards (MPEG-2, MPEG-4); second, the granularity of scalability is improved by SPEG combining temporal and SNR scalability. As shown in Figure 4.3, a priority (p_0 (highest)–p_{11}(lowest)) can be mapped to each layer. The QoS Mapper (see Figure 4.4, which depicts the SPEG pipeline and its components) uses the priority information to determine which layers are dropped and which are forwarded to the Net Streamer.

	I	B	P
Level 0	p_0	p_1	p_2
Level 1	p_3	p_4	p_5
Level 2	p_6	p_7	p_8
Level 3	p_9	p_{10}	p_{11}

Figure 4.3 SPEG layer model.

The decision to use SPEG as a layer-encoded format for the subjective assessment is based on the following reasons. SPEG is designed for a QoS-adaptive video-on-demand (VoD) approach, i.e., the data rate streamed to the client should be controlled by feedback from the network (e.g., congestion-control information). In addition, the developers of SPEG also implemented a join function that re-converts SPEG into MPEG-1 [131] allowing the use of standard MPEG-1 players, e.g., the Windows Media

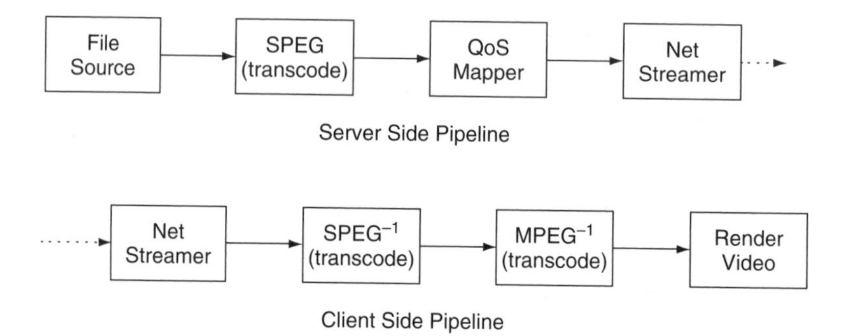

Figure 4.4 Pipeline for SPEG [131].

Player. Scalable video encoders available as products (e.g., [145, 146]) were not an option because videos created by those can only be streamed to the corresponding clients which allow neither the storage of the received data on a disk nor the creation of scheduled quality variations.

Figure 4.5 shows one single frame of an SPEG encoded sequence in its four possible steps of quality.

Base Layer

Base Layer + Layer 1

Base Layer + Layer 1 + Layer 2

Base Layer + Layer 1 + Layer 2 + Layer 3

Figure 4.5 One SPEG frame in four different quality levels (sequence Table Tennis).

4.3.2 Test Generation–Full Control

Since the test sequences must be created in a deterministic manner, the SPEG pipeline was slightly modified. The most important difference is: for the subjective assessment, data belonging to a certain layer must be dropped intentionally and not by an unpredictable feedback from the network or the client. This modification is necessary, since identical sequences must be presented to the test candidates in the kind of subjective assessment method that is used in the experiment. Therefore, the QoS Mapper is modified in such a way that layers are dropped at certain points in time specified by manually created input data. Additionally a second output path to the MPEG^{-1} module is added, making it possible to write the resulting MPEG-1 data in a file, and the NetStreamer modules are completely eliminated.

4.3.3 Measurement Method–Stimulus Comparison

The subjective assessment method is widely accepted for determining perceived image and video quality. Research that was performed under the ITU-R led to the development of a standard for such test methods [136]. The standard defines basically five different test methods which are briefly explained in the following. Figure 4.6 gives an overview of the classification of the different test methods. A detailed description of the single test methods can be found in reference [147].

- *DSCQE*: Double-stimulus continuous quality evaluation (DSCQE) is qualified for the assessment of new codecs or streaming video. The test is executed as follows: the original and the impaired video sequence are shown to the client in nondescript order. After the second sequence the test candidate assesses the quality of both sequences. Meanwhile, both sequences will be presented again and afterward the test candidate has to make the final assessment.

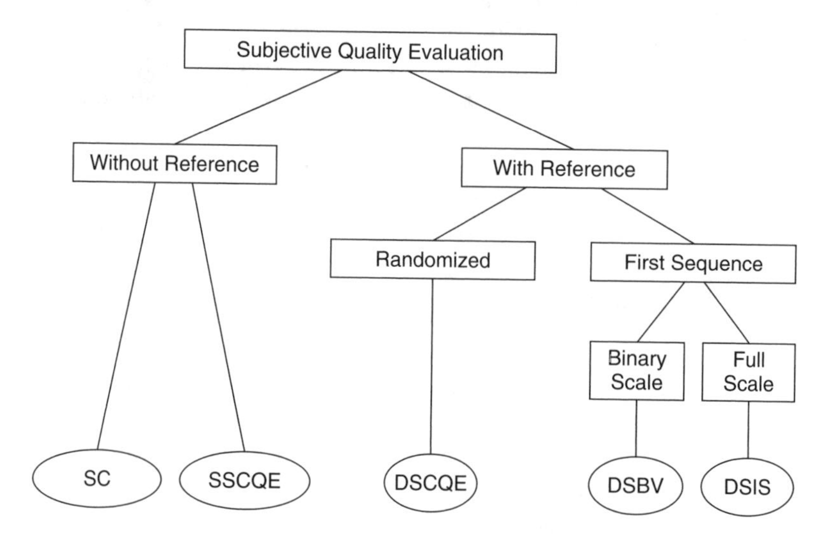

Figure 4.6 Test method classification.

- *DSIS and DSBV*: Two additional test methods which make use of reference sequences are double-stimulus impairment scale (DSIS) and double-stimulus binary vote (DSBV). The main difference between the two test methods is the comparison scale. While for DSIS a scale of five discrete values is used, the scale for DSBV is binary and allows only the assessment of whether the second sequence was equal to or worse than the first sequence. In both tests the reference sequence is always presented first.
- *SSCQE*: Single-stimulus continuous quality evaluation (SSCQE) is the first of two tests that belong to the category of test without a reference sequence. A sequence that can be of up to 30 minutes in duration is presented to the client. The quality of the sequence changes over time. The test candidate uses a control unit to assess the perceived quality of the sequence concurrently. The control unit is sampled every 2 to 5 seconds resulting in a graph that represents the video quality over time.
- *SC*: Stimulus comparison (SC) is another single-stimulus test method. With SC two impaired sequences are directly compared with each other. The test candidates do not assess the quality of a single sequence but the quality difference between the sequences. The assessment is executed, after the second sequence is shown to the test candidate, on a discrete scale consisting of seven options (see Table 4.1).

Since it is a major goal to investigate the basic assumptions about the quality of layer-encoded video, SSCQE is not the appropriate assessment method because comparisons between two video sequences are only possible on an identical time segment and not between certain intervals of the same video. In addition, SSCQE was designed to assess the quality of an encoder (e.g., MPEG-1) itself.

Table 4.1 Comparison scale

Value	Compare
−3	much worse
−2	worse
−1	slightly worse
0	the same
1	slightly better
2	better
3	much better

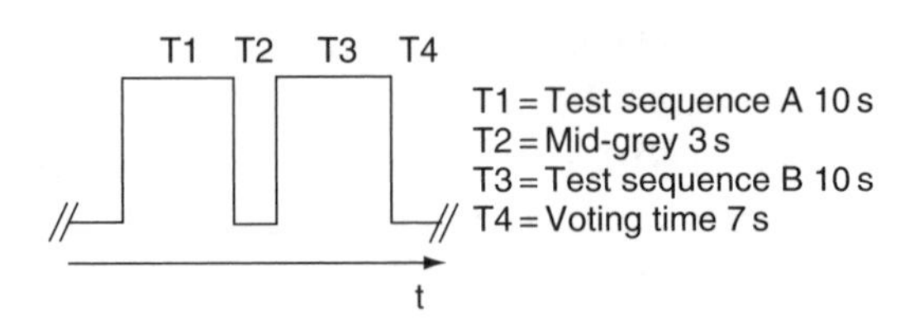

Figure 4.7 Presentation structure of test material.

Two test methods which better suit the kind of investigations performed in this work are DSCQS and DSIS. Compared to SSCQE they allow one to assess the quality of a codec in relation to data losses [134] and, therefore, are more suitable if the impairment caused by the transmission path is investigated.

The SC method differs from DSCQS and DSIS in that two test sequences with unequal qualities are shown (see Figure 4.7) and the test candidates can vote on a scale as shown in Table 4.1. Comparing two impaired videos directly with each other is the primary goal of this investigation. Since this is represented best by the SC method, the decision was made to use this method in the subjective assessment.

Additionally, preliminary tests (see Appendix B) have revealed that test candidates with experience in watching videos on a computer are less sensitive to impairment. That is, they recognize the impairment but do not judge it as annoying as do candidates who are not experienced. This effect is dampened since only impaired sequences have to be compared with each other in a single test that is based on the SC method. Preliminary tests with the DSIS method, where the original sequence and an impaired sequence are always compared, delivered results with less significance compared to tests performed with the SC method.

4.3.4 Test Application–Enforcing Time Constraints

Automated execution of the assessment was realized by an application [148] that was developed for this specific purpose. Since a computer is used to present the videos in any case, the candidates also perform their voting on the computer. Using this application has the advantage that time constraints demanded by the measurement method can be easily enforced, because voting is only possible during a certain time interval (exactly 7 seconds as shown in Figure 4.7). As a convenient side-effect, the voting data is available in a machine-readable format, thus simplifying the statistical analysis of the

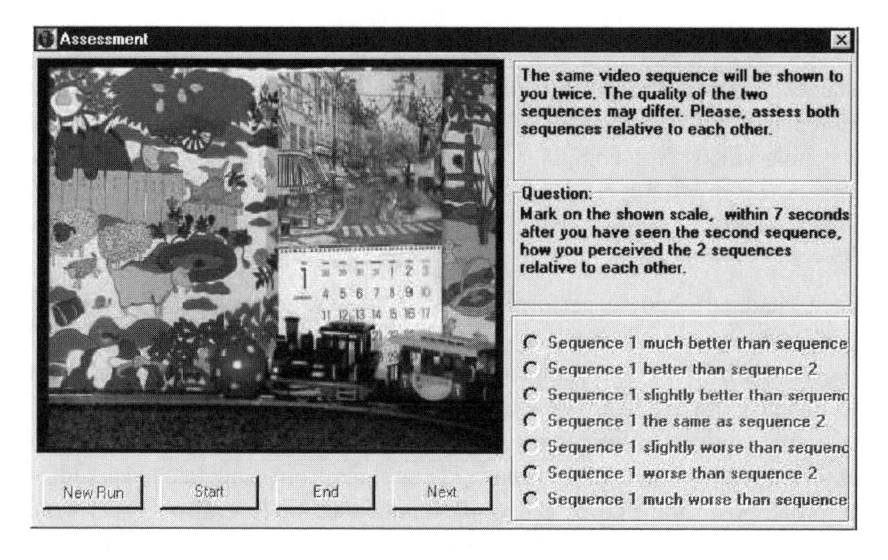

Figure 4.8 Test application.

gathered data. Figure 4.8 shows a snapshot of the application used to perform the subjective assessment.

4.4 EXPERIMENT

4.4.1 Scenario

Since quality metrics for layer-encoded video are very general, the focus had to be set on some basic test cases in order to keep the number of tests to be performed in the experiment feasible. Therefore, only isolated effects were investigated, one at a time, which on the one hand keeps the size of a test session reasonable and on the other hand still makes it possible to draw conclusions for the general assumptions, as discussed above. Thus, the investigation rather concentrates on observing the quality ranking for isolated effects such as frequency variations (as shown in sequences (b1) and (b2) in Figure 4.2) than on combined effects (as shown in Figure 4.1). This also has the advantage that standardized test methods [136], which limit the sequence length to several seconds, can be applied. All patterns that were used for the experiment are shown in Figure 4.10 and Figure 4.11.

4.4.2 Candidates

The experiment was performed with 115 test candidates (76 males and 39 females) between the ages of 14 and 64; 89 of them had experiences with watching videos on a computer.

4.4.3 Procedure

Each candidate had to perform 15 different assessments and each single test lasted for 33 seconds. All tests were executed according to the SC assessment method. The complete test session per candidate lasted for about 15 minutes,[†] on average. Three video sequences for this experiment that have been frequently used for subjective assessment [149] were chosen. The order of the 15 video sequences was changed randomly from candidate to candidate as proposed in the ITU-R B.500-10 standard [136] (see also Figure 4.9). After some initial questions (age, gender, profession) three assessments were executed as a warm-up phase. This should avoid the test candidates being distracted by the content of the video sequences, as reported by Aldridge

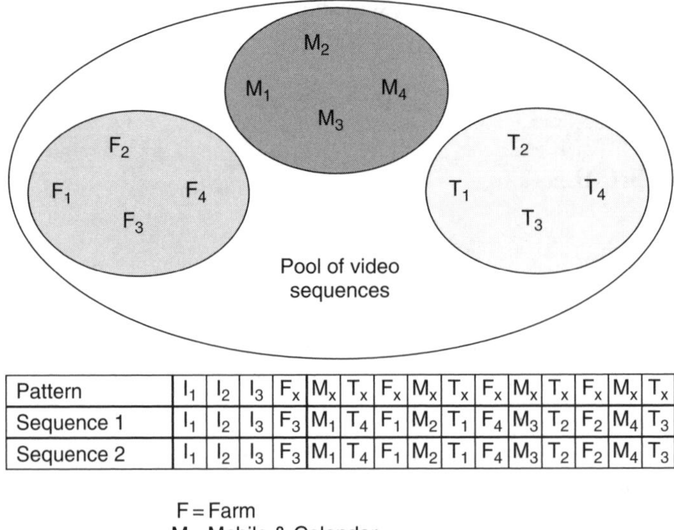

Pattern	I_1	I_2	I_3	F_x	M_x	T_x	F_x	M_x	T_x	F_x	M_x	T_x	F_x	M_x	T_x
Sequence 1	I_1	I_2	I_3	F_3	M_1	T_4	F_1	M_2	T_1	F_4	M_3	T_2	F_2	M_4	T_3
Sequence 2	I_1	I_2	I_3	F_3	M_1	T_4	F_1	M_2	T_1	F_4	M_3	T_2	F_2	M_4	T_3

Figure 4.9 Random generation of test sequence order.

[†] Watching the sequences and voting took less time, but the candidates had as much time as they wanted to read the questions and possible answers for each test ahead of each test.

et al. [137] (see also Appendix B). In order to avoid two consecutive video sequences (e.g., F_2 following F_1 immediately) having the same content, a pattern for the chronological order of the test sessions, as shown in Figure 4.9, is defined. F_x can be any video sequence from the F pool of sequences that has not been used in this specific test session, so far. Thus, a complete test session for a candidate could have the chronological order shown in Figure 4.9 (Sequence 1 and Sequence 2).

4.4.4 Layer Patterns

Figures 4.10 and 4.11 show the layer patterns of each single sequence that was used in the experiment, except for the first three warm-up tests where the comparison is performed between the first sequence that consists of four layers and the second that consists of only one layer. Each of the three groups shows the patterns that were used with one type of content. Comparisons were always performed between patterns that are shown in a row (e.g., (a1) and (a2)). As already mentioned in section 4.2, the goal of this investigation was to examine fundamental assumptions about the influence of layer changes on perceived quality. This is also reflected by the kind of patterns used in the experiment. It should be mentioned that the single layers are not equal in size (contrary to the presentation in Figures 4.10 and 4.11) as the size of the nth layer is given by the following expression: $s_n = 2s_{n-1}$ and, thus, segments of different layers have different sizes. Preliminary experiments showed that equal layer sizes are not appropriate to make layer changes perceivable; this is regarded as a realistic assumption since layered schemes that produce layers with sizes similar to the ones created by SPEG exist [150, 151].

In the experiment, a differentiation between two groups of tests is made, namely, one group in which the number of segments used by a pair of sequences is equal and one in which the number differs (the latter have a shaded background in Figures 4.10 and 4.11).

The main interest of the subjective assessment was in cases where it is necessary to compare the influence of additional segments that are added on different locations in a sequence (as shown in the test M&C1 in Figure 4.10). Results of those tests could have an implication on how to measure and improve the quality of existing retransmission scheduling techniques (see section 4.2.1). Yet, some of the effects such as the influence of the frequency of layer variations could only be investigated with test sequences consisting of a different number of segments (as shown in the test T-Tennis2 in Figure 4.11).

Since segments from different layers are not equal in size, the amount of data for the compared sequences differs. However, somewhat surprisingly,

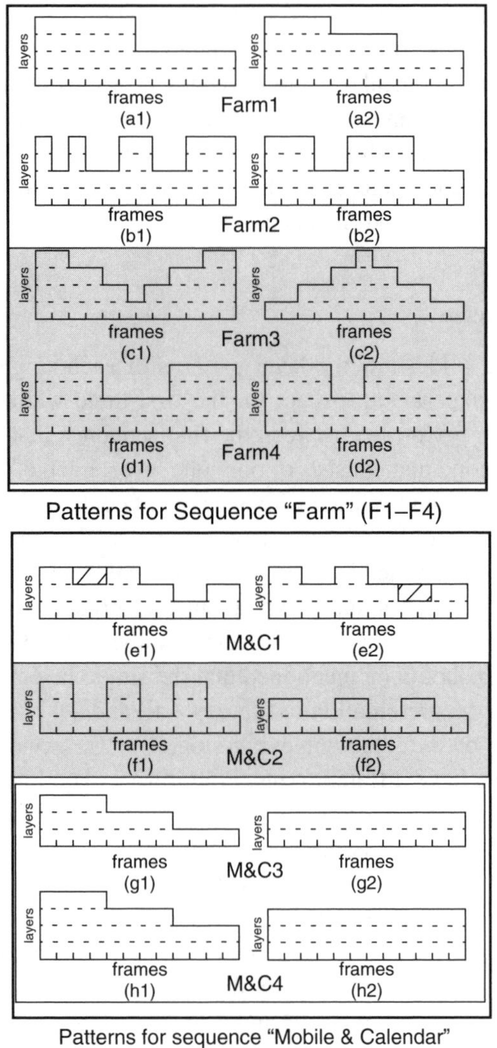

Figure 4.10 Segments that were compared in the experiment (Farm and M&C sequences).

and as discussed in section 4.5.3, a larger amount of data (resulting in a higher PSNR value) does not necessarily lead to a perception of higher quality. Additional tests with different quantities of segments between pairs were chosen to answer additional questions and make the experiment more consistent, as shown in section 4.5.2.

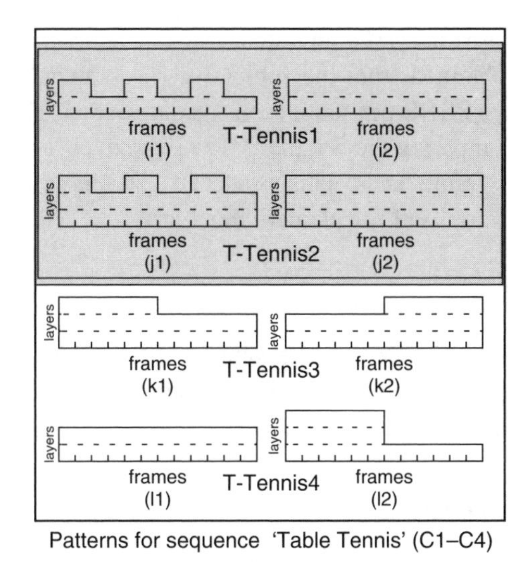

Figure 4.11 Segments that were compared in the experiment (T-Tennis sequences).

4.5 RESULTS

In the following, the results of the experiment described in section 4.4 are presented. Given the statistical nature of the data gathered it is obvious that the results presented cannot prove an assumption but only make it less or more likely. The overall results of all experiments are summarized in

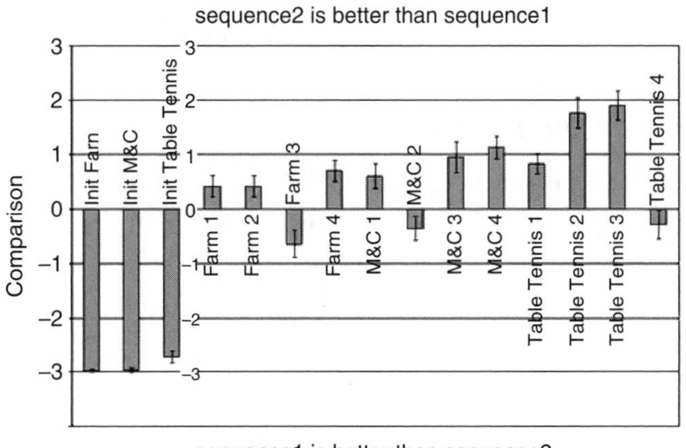

Figure 4.12 Average and 95% confidence interval for the different tests of the experiment.

Figure 4.12 and are discussed in the following subsections. Next to the statistical results obtained from the subjective assessment, objective data in terms of the average PSNR per sequence is also provided. The average PSNR was obtained by comparing the original MPEG-1 sequence with the impaired sequence on a per frame basis. This results in 250 single PSNR values per sequence, which were used to calculate the average PSNR.

4.5.1 Same Number of Segments

In this section, the results for the assessments of tests in which the total sum of segments is equal are discussed. That means that the space covered by the pattern of both sequences is identical.

4.5.1.1 Farm1: Amplitude

In this assessment the stepwise decrease was rated slightly better than one single but higher decrease (Figure 4.13). The result tends to justify the assumptions that were made about the amplitude of a layer change (as described in section 4.2.1).

4.5.1.2 Farm2: Frequency

The result of this test shows an even higher likelihood that the second sequence has a better perceived quality than in the case for Farm1. It tends to confirm the assumption that the frequency of layer changes influences the perceived quality, since, on average, test candidates ranked the quality of the sequence with fewer layer changes higher (Figure 4.14).

4.5.1.3 M&C1: Closing the Gap

This test tries to answer the question: Would it be better to close a gap in a layer on a higher or a lower level? The majority of the test candidates

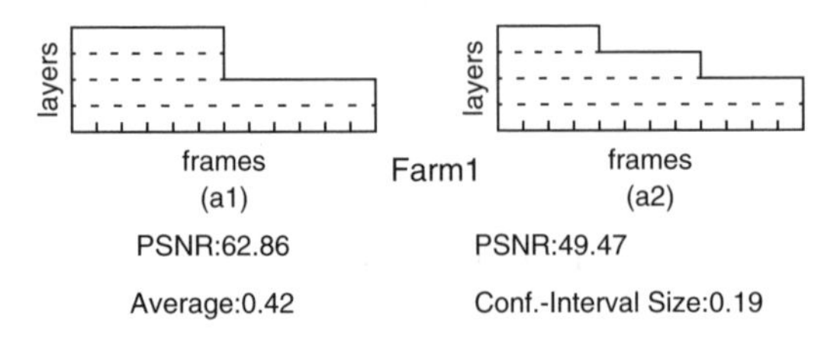

Figure 4.13 Farm1.

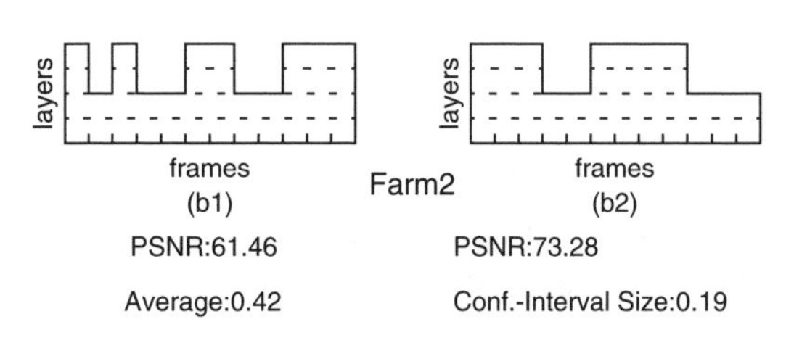

Figure 4.14 Farm2.

decided that filling the gap on a lower level results in a better quality than otherwise (Figure 4.15). This result tends to confirm assumptions made for retransmission scheduling (see Chapter 5).

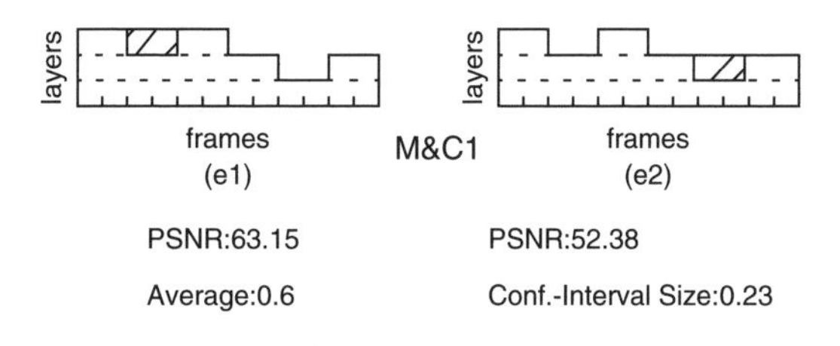

Figure 4.15 M&C1.

4.5.1.4 M&C3: Constancy

With an even higher significance than in the preceding tests, the candidates considered the sequence with no layer changes as the one with the better quality (Figure 4.16). One may judge this a trivial and unnecessary test, but the result is not that obvious, since (g1) starts with a higher number of layers. The outcome of this test implies that it might be better, in terms of perceived quality, to transmit fewer but a constant number of layers.

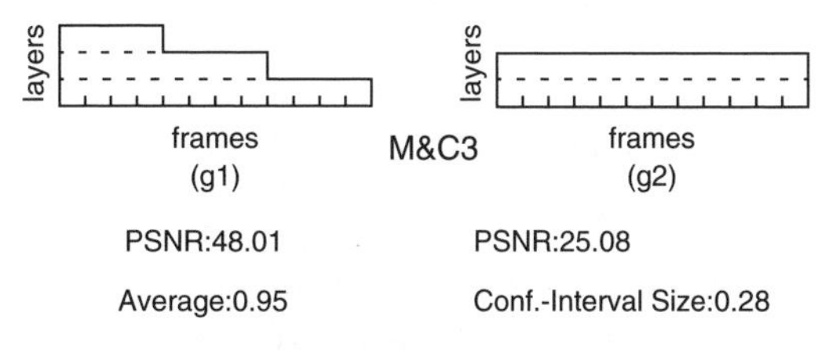

Figure 4.16 M&C3.

4.5.1.5 M&C4: Constancy at a Higher Level

This test was to examine whether an increase in the overall level (in this case by comparison to M&C3) has an influence on the perceived quality (Figure 4.17). Comparing the results of both tests (M&C3 and M&C4) shows no significant change in the test candidates' assessment: 66% of the test candidates judge the second sequences ((g2) and (h2)) of higher quality (values 1–3 in Table 4.1) in both cases which makes it likely that the overall level has no influence on the perceived quality.

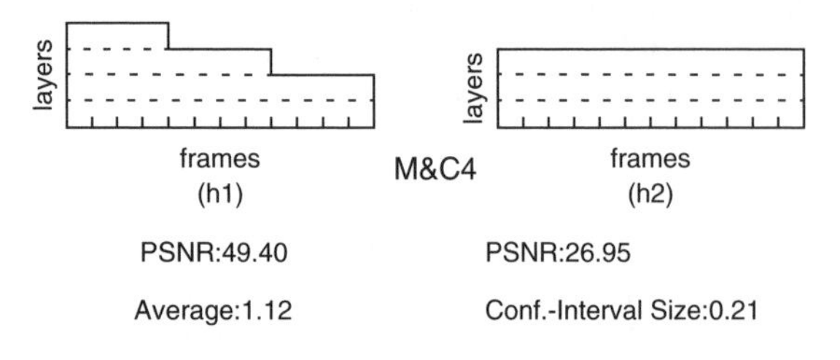

Figure 4.17 M&C4.

4.5.1.6 T-Tennis3: All is Well that Ends Well

The result of this test shows the tendency that increasing the number of layers in the end leads to a higher perceived quality (Figure 4.18). The result has a remarkably strong statistical significance (the highest bias of all tests).

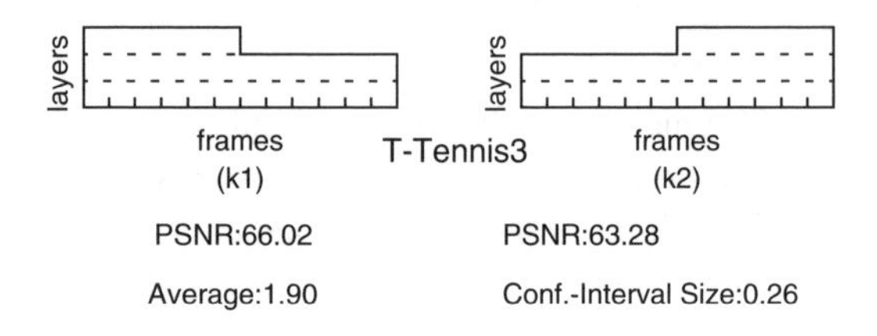

Figure 4.18 T-Tennis3.

4.5.1.7 T-Tennis4: The Exception Proves the Rule

The result of this test is a little bit surprising since it contradicts the results from Farm1 and M&C3 (Figure 4.19). It can only be assumed that the content might also have an influence on the perceived quality. But, to gain more insight in this phenomenon further experiments are necessary.

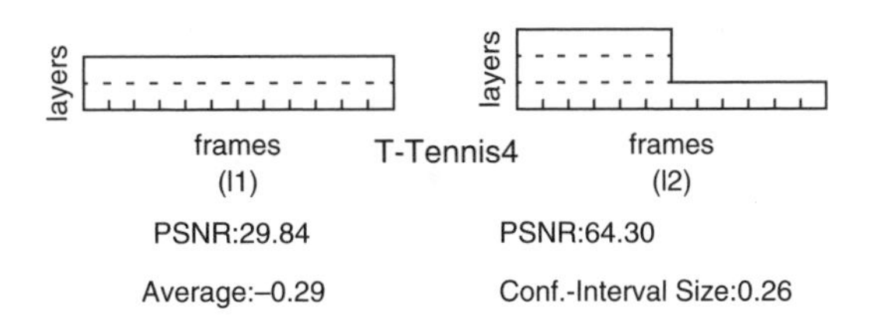

Figure 4.19 T-Tennis4.

4.5.2 Different Number of Segments

In the following five tests the total number of segments per sequence differs. All five tests have in common that the perceived quality of the sequences consisting of a pattern that covers a larger number of segments were ranked better. This is obvious, but it makes the overall result more consistent, because test candidates mostly realized this quality difference.

4.5.2.1 Farm3: Decrease vs. Increase

Starting with a higher number of layers, decreasing the number of layers, and increasing the number of layers in the end again seems to provide a better perceivable quality than starting with a low number of layers, increasing this number of layers, and going back to a low number of layers at the end of the sequence (Figure 4.20). This might be caused by the fact that test candidates are very concentrated in the beginning and the end of the sequence and that, in the first case details become clear right at the beginning of the sequence.

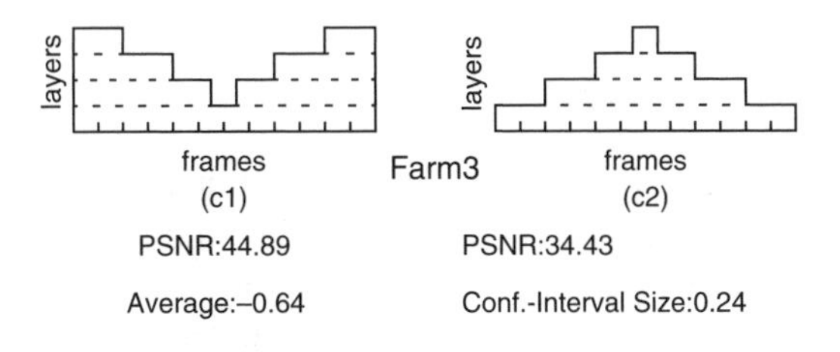

Figure 4.20 Farm3.

4.5.2.2 Farm4: Keep the Gap Small

In this test, the goal was to investigate how the size of a gap might influence the perceived quality (Figure 4.21). The majority of test candidates (66 out of 115) judged the quality of the sequence with a smaller gap slightly to be

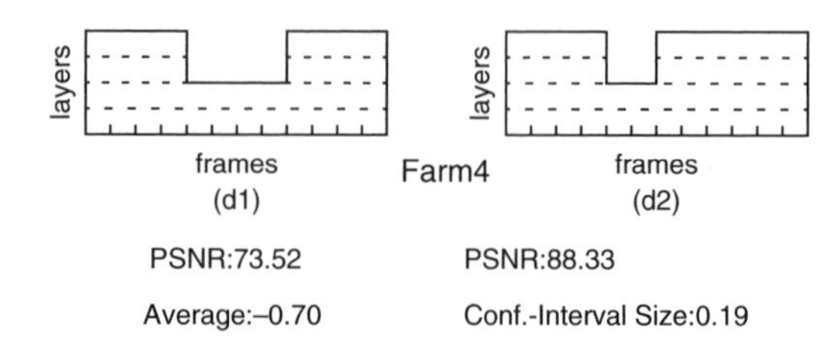

Figure 4.21 Farm4.

better. (Only 6 out of 115 judged the first sequence better.) This indicates that partly filling a gap can be beneficial.

4.5.2.3 M&C2: Increasing the Amplitude

The effect of the amplitude height was investigated in this test. The result shows that, in contrast to existing assumptions (see section 4.2.1), an increased amplitude can lead to a better perceived quality (Figure 4.22).

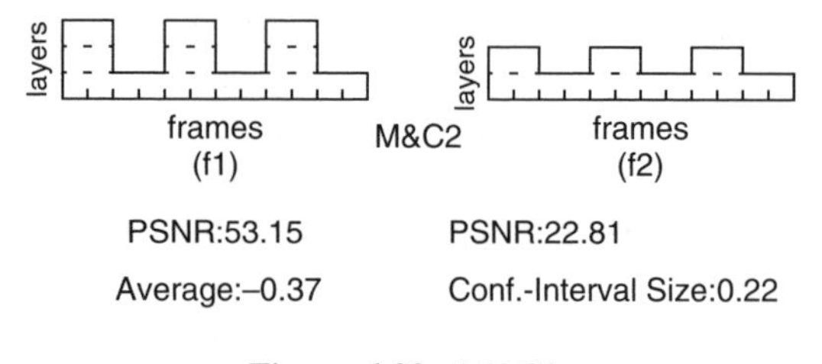

Figure 4.22 M&C2.

4.5.2.4 Tennis1: Closing All Gaps

This test is contrary to M&C2 in that the additional segments are used to close the existing gaps instead of increasing the amplitude of already better parts of the sequence (Figure 4.23). This strategy decreases the frequency of layer changes. Test candidates, on average, judged the sequence without layer changes better. The result of this test reaffirms the tendency that was already noticed in M&C1, that the perceived quality is influenced by the frequency of layer changes. The comparison of the results of M&C2 and T-Tennis1 show that a tendency towards filling the gaps and, thus, decreasing the frequency instead of increasing the amount of already increased parts of the sequence is recognizable. Definitely, further investigations are necessary to confirm this tendency, because here the results of tests with different

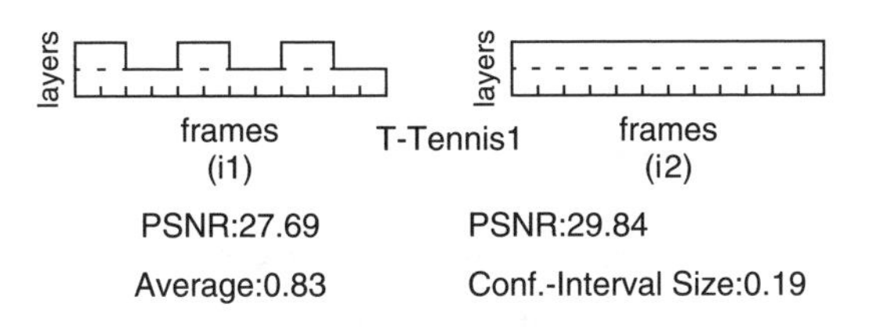

Figure 4.23 T-Tennis1.

contents are compared and the influence of the content on the perceived quality has not been investigated so far.

4.5.2.5 Tennis2: Closing All Gaps at a Higher Level

In comparison to T-Tennis1, here the interest is in how an overall increase of the layers (in this case by one layer) would influence the test candidates judgement (Figure 4.24). Again the sequence with no layer changes is judged better, even with a higher significance than for T-Tennis1. This might be caused by the fact that the number of layers is higher in general in T-Tennis2.

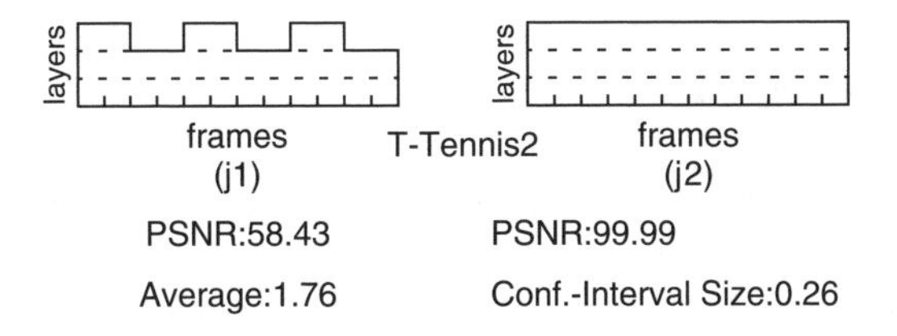

Figure 4.24 T-Tennis2.

4.5.3 Sequence Size and Quality

The PSNR is a popular metric to present the objective quality of video data. Therefore, the average PSNR of each sequence was also computed in order to investigate how subjective and objective quality are related. Since the determination of the objective quality can be performed with much less effort than a subjective assessment, the result of this investigation may provide hints as to whether the determination of the average PSNR is sufficient to define the quality of a video sequence. Note that, since the relation between subjective and objective quality is not the focus of the investigation presented in this section, this can only be seen as a by-product and would certainly need further investigation. (The PSNR values for each sequence are given in Figures 4.13 to 4.24.)

The results of the subjective assessments are contrary to the results of the PSNR in 8 of the 12 test cases. The results obtained for the test in which the sum of segments was equal for each sequence (section 4.5.1) are even stronger. They do not indicate a positive correlation between the two quality metrics (see Table 4.2). From the results of the subjective assessment we see a strong tendency that, in the case of layer-encoded video, the quality of a sequence is not well represented by the average PSNR.

Table 4.2 Comparison between subjective and objective quality (same number of segments)

Shape	Farm1	Farm2	M&C1	M&C3	M&C4	T-T3	T-T4
PSNR of shape 1	62.86	61.46	63.15	48.01	49.40	66.02	29.84
PSNR of shape 2	49.47	73.28	52.38	25.08	26.95	63.28	64.30
Average of assessment	0.42	0.42	0.60	0.95	1.12	1.90	−0.29

contrary to subjective assessment in accordance with subjective assessment

4.6 THE SPECTRUM

The investigation presented in section 4.5.3 reveals that the average PSNR is not well suited to represent the perceived quality of layer-encoded video. This lack of an appropriate objective quality metric for layer-encoded video leads to a new objective metric called the *spectrum*. The goal during the development of the spectrum was to express the factors that influence the perceived quality (as presented in section 4.5) through a mathematical expression giving similar results to those obtained by the subjective assessment.

Therefore, the *spectrum* of a cached layer-encoded video, v, can be introduced:

$$s(v) = \sum_{t=1}^{T} z_t \left(h_t - \frac{1}{\sum_{i=1}^{T} z_i} \left(\sum_{j=1}^{T} Z_j h_j \right) \right)^2 \tag{2}$$

with h_t and z_t defined as:

- h_t, number of layers in time slot t, $t = 1, \ldots, T$
- z_t, indication of a step in time slot t, $z_t \in \{0, 1\}$, $t = 1, \ldots, T$

Without loss of generality, a slotted time period with slots corresponding to the transmission time of a single (fixed-size) segment is assumed as well as that all layers are of the same size.

The spectrum captures the frequency as well as the amplitude of quality variations. The amplitude is captured by the differences between quality levels and average quality levels where larger amplitudes are given higher weight owing to squaring these differences. The frequency of variations is captured by z_t. Only those differences that correspond to a step in the cached layer-encoded video are taken into account. While the spectrum as defined in equation (2) looks very similar to the usual variance of quality levels for the cached video, it is important to note that the introduction of the z_t takes into account the frequency of changes in the quality levels. A spectrum of the value 0 represents the best possible quality, while the spectrum increases with a decreasing quality. An example calculation of the spectrum is shown in Figure 4.25.

The decision to name this objective quality metric spectrum is caused by the fact that it is influenced by the amount of layer variation (i.e. the frequency of layer variation). It does not reflect what is meant by the spectrum in a strong mathematical sense (the result of the transformation from the time domain into the frequency domain) and, therefore, must not be mistaken with it.

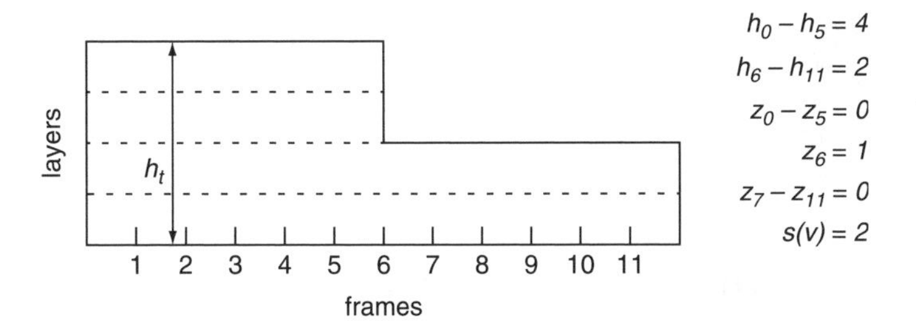

Figure 4.25 Parameters to calculate the spectrum.

4.6.1 Comparison of the Spectrum with the Subjective Assessment Results and the PSNR

In Table 4.3, the spectrum of the shapes presented in section 4.3.1 is compared with the results from the subjective assessment and the PSNR for each sequence. With the exception of the cases for the Farm1 (Figure 4.13) and T-Tennis3 (Figure 4.18) tests, there is a consistency between the subjective quality and the spectrum. In contrast to the subjective results, the PSNRs of the sequences are only consistent in one of the six cases (test Farm2). This argues for the hypothesis that the spectrum is more suitable as an objective quality metric than the PSNR because the PSNR of each shape and the results of the subjective assessment are not consistent, as shown in section 4.5.3. But it must be mentioned that the spectrum does not regard time dependencies very well, as can be seen in the case of the T-Tennis3 test (Figure 4.18) where the spectrum is equal for both cases but the second shape was assessed as having a higher quality. The latter is due to the fact that the number of layers increases towards the end of the sequence. With the knowledge that the spectrum is a suitable metric for the quality of layer-encoded video, it was chosen to rate and compare the retransmission scheduling algorithms that are presented in Chapter 5.

Table 4.3 Comparison among spectrum, subjective and objective quality

Shape	Farm1	Farm2	M&C1	M&C3	M&C4	T-Tennis3
$s(v)$ of shape 1	2	6.86	2	2	2	0.5
$s(v)$ of shape 2	2	4	1	0	0	0.5
PSNR of shape 1	62.86	61.46	63.15	48.01	49.40	66.02
PSNR of shape 2	49.47	73.28	52.38	25.08	26.95	63.28
Average of assessment	0.35	0.55	0.73	1.18	1.02	2.18

contrary to subjective assessment

in accordance with subjective assessment

inconclusive

4.7 IMPLICATIONS FOR MDC AND FGS

Section 4.3.1 explains why, for the initial investigation of layer changes on the viewer's perceived quality, a discrete hierarchical layer encoded format was chosen. Once again, the main aim of this investigation was to learn how retransmission scheduling can be influenced by understanding how layer variations influence the perceived quality of a video. Since such an investigation has not been performed, it was the goal to answer some fundamental questions which also can be seen as trends for related layer-encoding formats. To obtain these answers the experiment presented in this chapter was the most appropriate. Nevertheless, the results from the experiment can be seen as trends for MDC and FGS layer-encoded video, as will be discussed in the following.

4.7.1 MDC

Since MDC layers have the same discrete characteristics as the SPEG format, the results for an MDC-based experiment are expected to be quite similar to those presented here. As mentioned in section 3.3.3, the main advantage of MDC is its non-hierarchical nature that makes it very suitable for cases where a transmission via diverse paths is possible or necessary.

Figure 4.26 gives an example that can only occur with MDC. It is also shown how retransmissions for segments of missing layers would be performed on the basis of the results of the the experiment presented in this chapter. It must be mentioned that, although applying the results obtained to an MDC scheme seems to be straightforward, a subjective assessment for the MDC case must be performed to prove the correctness of these assumptions.

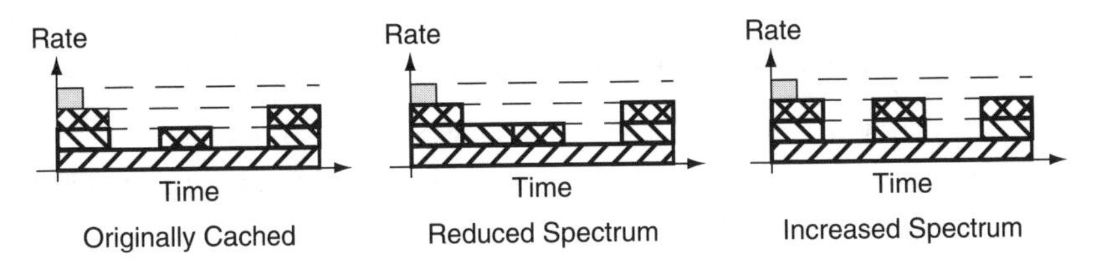

Figure 4.26 Retransmissions for MDC.

4.7.2 FGS

Adapting the rate of an FGS-encoded video to each rate change the transmission protocol experiences might be annoying for the viewer. Although TCP-friendly protocols specificially designed for the purpose of streaming try to reduce the variation in transmission, the frequency of this changes is still high (see Chapter 7). With FGS this can lead to a large amount of quality changes. As shown by the results in this chapter a large amount of quality change can reduce the perceived quality of a video.

To avoid this high frequency of bandwidth changes in the enhancement layer one could apply a more coarse-grained transmission scheme. That means the sender would only increase the the transmission bandwidth for the enhancement layer if the available transmission rate exceeds a certain limit. This would eventually lead to a transmission scheme very similar to the one for discrete hierarchical layer-encoded video, as shown in Figure 4.27. Reference [120] provides one example where FGS is used in such a thresholded way in combination with a distribution infrastructure that contains caches. If FGS is used in a discrete form, as shown in Figure 4.27, the results from

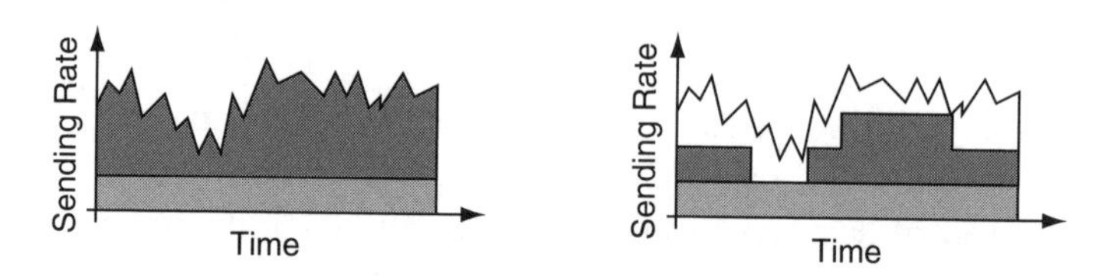

Figure 4.27 FGS vs. thresholded FGS.

the subjective assessment presented in this chapter should be applicable for the retransmission of missing parts of the enhancement layer.

4.8 SUMMARY

A statistical analysis of the experiment mostly validates assumptions that were made in relation to layer variations and the perceived quality of a video:

- The frequency of variations should be kept as low as possible.
- If a variation cannot be avoided, the amplitude of the variation should be kept as small as possible.

One basic conclusion from the results in section 4.5.2 is: adding information to a layered video increases its average quality. But adding information at different locations can have a substantial effect on the perceived quality. Assumptions made for heuristics in retransmission scheduling could be substantiated by this investigation. That means it is more likely that the perceived quality of a layer-encoded video is improved if

- the lowest quality level is increased, and
- gaps in lower layers are filled.

The results from section 4.5.3 should be used to refine the retransmission scheduling heuristics in relation to the size of each single layer. Therefore, the metric that represents the quality improvement must also take into account that it might be more expensive to retransmit a segment of layer $n+1$ than of layer n. Another interesting outcome of the experiment is the fact that a quality improvement may be achieved by retransmitting less data, if a layered encoding scheme is used in which the layers are not of identical size. The results obtained can, in addition, be used to refine caching replacement policies that operate on a layer level [115] as well as layered multicast transmission schemes which try to offer heterogeneous services to different subscribers, as for example in the receiver-driven layered multicast RLM [100] scheme and its derivations.

The results of this investigation clearly strengthen the assumption that a differentiation between objective and subjective quality, in the case of variations in layer-encoded video, must be made, if the objective measure is based on the PSNR. A new objective quality measure (the spectrum) that was developed on the basis of the results of the subjective assessment is a more suitable metric for the quality of layer-encoded video compared to the PSNR.

5

Retransmission Scheduling

5.1 MOTIVATION

With SAS, it is very likely that videos are not cached in their best quality when they are cached for the first time. However, for subsequent requests which shall be served from the cache it may be unattractive to suffer from the possibly very bad or strongly varying quality experienced by the initial transmission of the video that has been selected for caching. Therefore, missing segments of the cached video should be retransmitted to enable higher quality service from the cache to its clients. The most interesting issue here is how to schedule the retransmissions, i.e., in which order to retransmit missing segments, in order to achieve certain quality goals for the cached video content. A further design issue is, when to schedule retransmissions.

The method of retransmission scheduling can also be used in clients, if they allow buffering and the buffer can store at least as much data as can be sent during an RTT [97]. However, the focus in the SAS architecture is on retransmission into caches, since these retransmissions can be beneficial for more than one client and caches are a necessary element of *scalable adaptive streaming* in any case. This mechanism is called retransmission owing to the fact that an entire layer was not transmitted for a certain interval due to congestion on an intermediate link, but would have been transmitted if no congestion had occurred. Figure 5.1 gives a basic overview of the retransmission process. A video object is initially cached as shown by the grey-coloured shape. At a later point in time this object is requested by a client which is also served by the cache. The cache now requests missing segments and forwards them in combination with the segments of the cached video object to the client. Thus, the quality of the video object streamed to

---

Scalable Video on Demand: Adaptive Internet-based Distribution M. Zink

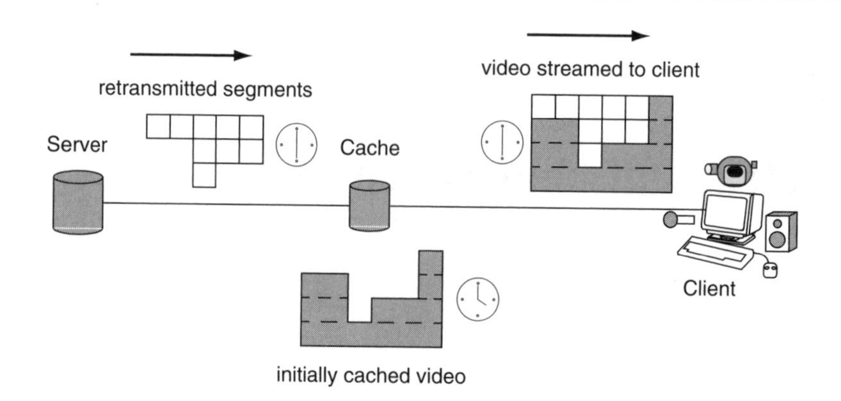

Figure 5.1 Retransmission scheduling.

the client has a higher quality than the initially cached object. Depending on the decision of the cache, the retransmitted segments may also be stored on the cache in order to increase the quality of the cached object. When and how these retransmissions are performed is discussed in more detail in section 5.1.1.

In this chapter, an investigation on retransmission scheduling is described. The goal of this investigation is to develop mechanisms for retransmission scheduling to improve the quality of layer-encoded video at client and cache. Since an investigation on optimal retransmission scheduling showed that it cannot be applied in SAS because it is computationally infeasible (see section 5.2), heuristics are developed which have a much lower complexity. The creation of new heuristics is also prompted by the fact that an existing heuristic for retransmission scheduling revealed some drawbacks, as is shown in section 5.4. Knowledge gained by the subjective assessment (see Chapter 4) is used for the creation of the new retransmission scheduling heuristics. A performance analysis of these heuristics is carried out through simulations (see section 5.5). Following the retransmission scheduling heuristics four different approaches that focus on the location of the maximum quality improvement are presented. During the investigation on retransmission scheduling it became obvious that retransmissions can be performed to maximize the quality of a layer-encoded video at either the cache (see section 5.6) or the client (see section 5.4). Thus simulations for the two approaches are performed which lead to a third, hybrid, approach which maximizes the quality for the current viewer while increasing the quality on the cache almost to the maximum (see section 5.7).

5.1.1 Retransmission Time

There are four possible occasions when retransmissions can be performed. All four have the goal to improve the quality of the cached content as soon as possible after the decision has been made to start the retransmission:

(1) Directly after the initial streaming process: the cache starts requesting missing segments without waiting for further requests for a certain video. This allows one to offer the highest possible quality to requests that arrive during the retransmission phase.
(2) During the initial streaming process: the cache starts requesting retransmissions a certain amount of time after the start of the streaming session. Data that is retransmitted does not improve the quality for the actual viewer, but increases the quality of the cached video object and, thus, the quality for viewers that request the object at a later point in time.
(3) During subsequent requests: the cache serves subsequent requests but, simultaneously, also orders missing segments from the server.
(4) During requests for different content from the server: in reference [92] and Chapter 7, a technique is presented that allows the transmission of requested segments (for an already cached video) in addition to video data that is streamed from the server to the cache.

The first and second alternative have in common that a cached video's quality is improved as quickly as possible, with the difference that in version (1) an independent session is created to perform the retransmissions.

The third alternative inherits the advantage of write-through caching that any bandwidth between the cache and server is used only if a client request is directly related to it. This is a major advantage in environments where bandwidth between server and cache is scarce. With the fourth alternative, the cache can decide about which of the cached video retransmissions should be performed when a new video is requested from the server. All four can be supported by retransmission scheduling, since it only determines in which order missing segments should be sent to the cache. The importance of retransmission scheduling is affirmed by the results from section 5.5.2 ('number of layers') which show that, depending on the number of layers a video consists of, one retransmission phase might not be sufficient to improve the quality of the cached video to its maximum level.

Figure 5.2 gives an example for the four versions of when to perform retransmissions. The reason that only four segments are retransmitted in the case of versions 3 and 4 is that it is assumed that the additional available bandwidth for retransmissions was equivalent to four segments. This is not the case for version 1 where an independent session is created to transmit

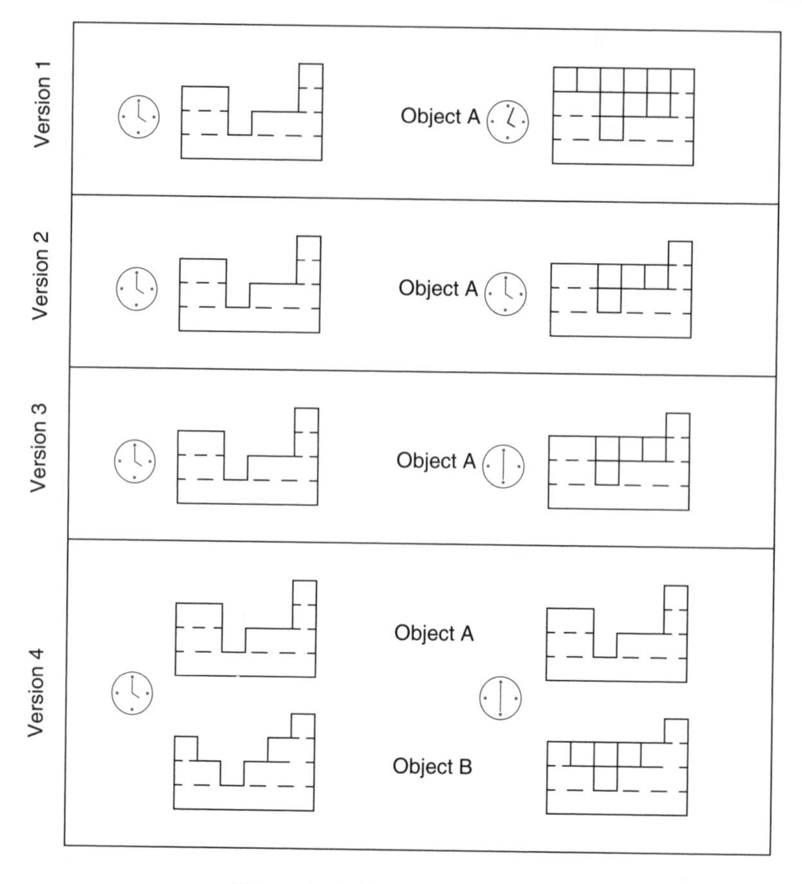

Figure 5.2 Retransmission time.

missing segments from the server to the cache and the transmission time is not limited.

Let us recall the example from section 2.6 where students from different universities use the distribution system to receive on-line lectures from remote universities. In this case, the four possibilities of performing retransmissions could be as follows:

- The popularity information of the requested object is known, and based on this information further requests for this specific object are very likely. This might be the case for lectures advanced in time where popularity information from preceding sessions of this lecture reveal that it is very likely further clients served by this cache might also request this object. If the collected popularity data also allowed one to draw conclusions about the time the requests for a single object had been made, and this data showed requests following each other in a time period shorter than the playout duration of the video object, applying the first of the four retransmission options would be a good choice. The quality of the cached video would be increased as soon as possible and clients requesting the object during the retransmission phase would receive the highest possible quality.

- An option with the method above is to perform the retransmission during the actual streaming phase. This can, for example, be performed with the fair share claiming (FSC) method, as presented in Chapter 7. The drawback of this method is the fact that probably not all missing segments are retransmitted and the video object is not cached in its highest quality.
- The popularity information about the requested video object may not be known and, therefore, it may not be foreseeable whether the object will ever be requested again. In this case, it is not assured that retransmissions will be of any benefit. Thus, retransmissions are only performed if further requests for this object are made. For example, when the first session of a lecture is requested for the first time, no popularity information about this video object belonging to this lecture is available on the cache serving the client posing the request.
- Owing to bandwidth limitations not all requested segments may be transmitted in a retransmission session from the server to the cache. Yet, popularity information shows a high probability for further requests of this object. For example, although retransmission scheduling was performed, a session of a very popular lecture (represented by video object A) is not cached in full quality. A client requests a different lecture (video object B) with low or unknown popularity. In this case, retransmission scheduling can be performed for video object A while the streaming phase of video object B is taking place.

5.1.2 Retransmission Focus

Having considered the retransmission time, one can also differentiate between a retransmission scheduling approach which has the goal of maximizing the quality for the current viewer and an approach that maximizes the quality of the cached video. The first approach is described as *viewer-centric*, while the second is described as *cache-centric*. The viewer-centric approach will be presented in more detail in section 5.4 and the cache-centric approach in section 5.6. A modified version of the viewer-centric approach called *cache-friendly viewer-centric* is presented in section 5.7.

5.1.3 Scheduling Goals

The rationale for making an effort to schedule retransmissions in an intelligent way is that the presentation quality for users that are served from the cache can be enhanced. Therefore, it has to be made explicit what constitutes a quality enhancement, i.e., a goal for retransmission scheduling algorithms to strive for is needed. However, it is commonly assumed and also shown in Chapter 4 that users react very sensitively to quality variations of a video [101]. Hence, a retransmission scheduling algorithm that tries to avoid or even decrease quality variations for a cached video can be considered superior to others which do not take this into account. The results of the subjective assessment performed in Chapter 4 show that the negative effect of quality variations has two dimensions: the frequency of variations and the amplitude of variations. The goal of retransmission scheduling should be to minimize both. In Chapter 4, a new subjective quality metric called the spectrum is presented that can be used to determine the quality of a layer-encoded video. Since the quality of a layer-encoded video is inversely proportional to the resulting value of the spectrum, the retransmission scheduling goal for a video, v, can now be stated as the minimization of the spectrum $s(v)$.

5.2 OPTIMAL RETRANSMISSION SCHEDULING

In this section, resolving the complexity of retransmission scheduling by looking at optimal retransmission scheduling is discussed. Since the determination of optimal retransmission schedules is either computationally infeasible or at least intensive, some heuristic schemes are presented. One of them has been proposed in reference [115], whereas the others have been newly developed in response to the shortcomings of the former.

The goal of retransmission scheduling is to minimize the spectrum of an already cached layered video subject to the constraint that any available bandwidth is used for retransmissions. This constraint ensures that a cached video is further enhanced even if for all time slots the same quality level is reached, i.e., the spectrum equals 0.

A formulation of optimal retransmission scheduling as a mathematical program is given in Figure 5.3. Here, the overall available retransmission

d_t	–	number of retransmitted layers for time slot t
h_t	–	number of layers in time slot t
ν'	–	the cached video after retransmissions
H	–	the maximum number of layers
$\tilde{B}(\bar{t})$	–	estimated amount of overall retransmission capacity for all time slots till $\bar{t}$

$$\text{Minimize } s(\nu') \quad (3)$$

$$\text{subject to} \quad (4)$$

$$\sum_{t=1}^{\bar{t}} d_t = (\tilde{t})\bar{B} \qquad \forall \bar{t} = 1, \ldots, T \quad (5)$$

$$h_t + d_t - h_{t-1} - d_{t-1} \leq Hz_t \qquad \forall t = 1, \ldots, T \quad (6)$$

$$h_{t-1} + d_{t-1} - h_t + d_t \leq Hz_t \qquad \forall t = 1, \ldots, T \quad (7)$$

$$H - h_t \geq d_t \geq 0 \qquad \forall t = 1, \ldots, T \quad (8)$$

Figure 5.3 Optimal retransmission scheduling model.

capacity is modelled as an estimate. Yet, in this investigation a constantly available bandwidth is assumed, i.e.,

$$\tilde{B}(\bar{t}) = \frac{B}{T} \times \bar{t} \quad (9)$$

where B is the overall retransmission capacity for the video. This is certainly a simplifying assumption; but the algorithms presented in the following do not depend on it. Simulations with a varying bandwidth presented in Chapter 7 did not show any significant influence on the algorithm's performance.

Optimal retransmission scheduling is a discrete nonlinear optimization problem. As such it is, to the best of our knowledge, analytically intractable. It is very similar to the quadratic assignment problem, which is known to be NP-complete [152]. The following example is given to illustrate the complexity of retransmission scheduling. Considering the search space for an exhaustive search, assuming that in each time slot at least one layer is missing, then a *lower* bound for the size of the search space can be obtained:

$$\binom{T}{B}$$

Table 5.1 Execution times for optimal retransmission scheduling

Segments to retransmit	1	2	3	4	5	6	7
Optimal Duration (seconds)	0.0003	0.0054	0.1222	2.8206	56.1878	1168.5	23686.5
Optimal Spectrum	20.73	18.92	16.73	16.4	14.55	14.1	12.55
U-SG-LLF Duration (seconds)	5.3×10^{-5}	5.4×10^{-5}	5.4×10^{-5}	5.4×10^{-5}	5.5×10^{-5}	5.5×10^{-5}	5.5×10^{-5}
U-SG-LLF Spectrum	20.73	22.67	16.73	18.92	18.92	18.92	12.55

For example, for 100 time slots and a retransmission capacity of 50 this amounts to 1.534×10^{93} possible ways of reordering missing segments (and this is only a very loose lower bound). Thus, given reasonable restrictions on computing power, an exhaustive search for reasonable values of the number of time slots T is computationally infeasible.

Next to the heuristics that are presented in the following, the optimal retransmission scheduling algorithm, based on an exhaustive search, was also implemented in the custom simulation environment that is described in section 5.5. To get an impression of how long it would take to determine the optimal set of missing segments that should be retransmitted the execution time for the algorithm was measured for the following example. An initially cached video, similar to the one shown in Figure 5.4, is used as the starting point for the algorithm. Table 5.1 shows the time that it took the algorithm to calculate the optimal set of segments for retransmission starting from 1 up to 7 possible segments. The measurement was executed on a standard PC running Linux RedHat 7.3 with a Pentium III (500 MHz) and 500 Mbytes of main memory.

The results of this measurement reveal that the optimal algorithm cannot be applied for retransmission scheduling, especially with the knowledge that the retransmission schedule, in some cases, must be calculated on the order of seconds. In the case of seven segments that can be retransmitted the resulting spectrum that is obtained by the optimal algorithm is identical to the one that is obtained by the application of a user-centric unrestricted heuristic (see Figures 5.5, 5.7 and 5.8). The amount of time needed to determine the schedule with the algorithm that is based on the unrestricted heuristic is negligible compared to the optimal algorithm. The resulting values for the spectrum shown in Table 5.1 demonstrate the drawback of the heuristic. The optimal spectrum will not be determined in all cases, but, considering the performance gain, the heuristics are an applicable alternative.

5.3 HEURISTICS FOR RETRANSMISSION SCHEDULING

There are two basic approaches in retransmission scheduling that influence the design of the heuristics developed for this purpose. The first approach is to maximize the quality for the current viewer, while the second is to improve the overall quality of the cached video object. In section 5.4, heuristics for the first approach are presented and in section 5.6 it is shown how a simple modification allows the application of these heuristics for the second approach. Section 5.7 presents a combination of the two aforementioned approaches.

5.4 VIEWER-CENTRIC RETRANSMISSION SCHEDULING

As mentioned above, retransmission algorithms are intended to improve the received quality of the current viewer or the quality of the stored version on the cache. Here, the focus is on improving the quality for the current viewer at the client. Thus, it must be assured that segments requested for retransmission arrive at the client before or at the actual playout deadline. All of the presented heuristics take this requirement into account, but it is also shown that this is the limiting factor in terms of obtaining an identical copy of the video object on the cache.

5.4.1 Window-based Lowest Layer First (W-LLF)

The first heuristic that is investigated is proposed in reference [115]. It is fairly simple and called 'Window-based lowest layer first' (W-LLF), because the cache always looks a certain number of time slots ahead of the current playout time and requests retransmissions of missing segments from the server in ascending order of their layer levels. To ensure that the retransmitted segments do not arrive after their playout time (t_p) to the current client, a prefetching offset, O_p, for the examined time window is introduced. O_p should be chosen sufficiently large such that $O_p > \text{RTT}$ for the transmission path between server and cache at all times. Overall, the time window $[t_p + O_p, t_p + O_p + W]$ slides over the video in discrete steps of length W until it is finished (t_{end}). The operation of the algorithm is further illustrated in Figure 5.4.

W-LLF has some obvious disadvantages:

- If, for example, an already complete area (all layers are entirely cached) is scanned, no retransmissions are scheduled for this prefetching window, although there may very well be later parts of the video which could benefit from retransmissions.
- It is possible that the currently available bandwidth between server and cache would allow the transmission of more segments than those that are missing in the current prefetching window and again additional segments could be requested from the server to allow for a faster quality upgrade of the cached video.

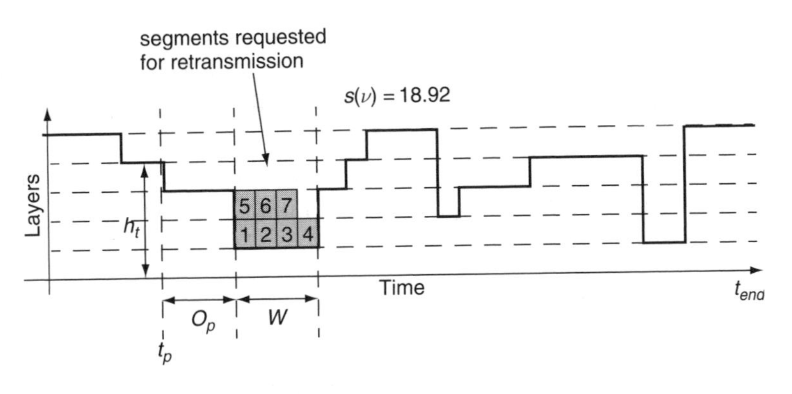

Figure 5.4 W-LLF operation.

Although, these obvious drawbacks might be eliminated by extensions of the W-LLF algorithm, they exhibit a fundamental weakness of W-LLF: the restriction of scheduling missing segments for retransmission only for a certain number of time slots ahead. Therefore, W-LLF is likely to be rather shortsighted with respect to the scheduling goal of minimizing the spectrum of videos stored on the cache. In the following, a new kind of retransmission scheduling algorithm that eliminates this restricted view is introduced.

One rare case in which W-LLF can be superior compared to the heuristics presented in the following is when the viewer decides to not watch the video completely. Then, segments requested for retransmission that are temporally located after the moment the viewer decides to stop watching the video do not improve its perceived quality. This situation can occur with the unrestricted heuristics presented in the following but has no influence on W-LLF if the window size is kept relatively small.

5.4.2 Unrestricted Priority-based Heuristics

The problems with W-LLF as described above lead to the idea of avoiding the use of a prefetching window, W, for retransmission scheduling. That means an unrestricted look at all missing segments ahead of the current playout time t_p (plus the prefetching offset O_p) is taken when making requests for retransmissions from the server. Note that the unrestricted algorithms still send periodic retransmission requests to the server (every W time slots) to ensure on the one hand that retransmissions and playout to the client are kept synchronized and on the other hand that the modified shape of the cached video due to retransmitted segments can be taken into account by the scheduling algorithms.

A further goal was to adapt the scheduling decisions to the results obtained from the subjective assessment (see Chapter 4), instead of rigidly assigning the highest priority to the first segments with the lowest layer level, as performed by previous approaches [115].

In the following, three heuristics of the more general class of unrestricted priority-based retransmission scheduling algorithms are described.

5.4.2.1 Unrestricted Lowest Layer First (U-LLF)

This algorithm is very similar to W-LLF because it uses as priority solely the layer level. In contrast to W-LLF, however, it always scans the interval $[t_p + O_p, t_{end}]$ in order to request missing segments from the server (every W time slots). Figure 5.5 gives an example of how the missing segments would be scheduled for retransmission. The numbers for each segment define the order in which the segments should be sent from the server.

5.4.2.2 Unrestricted Shortest Gap Lowest Layer First (U-SG-LLF)

Considering the definition of the spectrum in section 4.6 and taking into account the scheduling goal of minimizing the spectrum, one can observe that there are, in principle, two ways to decrease the spectrum of a video: to increase the lowest quality levels (which is taken care of by choosing the lowest layer levels first), *or* to close gaps in the video, i.e., reduce the number of $z_t \neq 0$. The latter is not captured by simply using layer levels as priorities. Figure 5.6 gives an illustrative example of how the spectrum is affected by the closing of gaps.

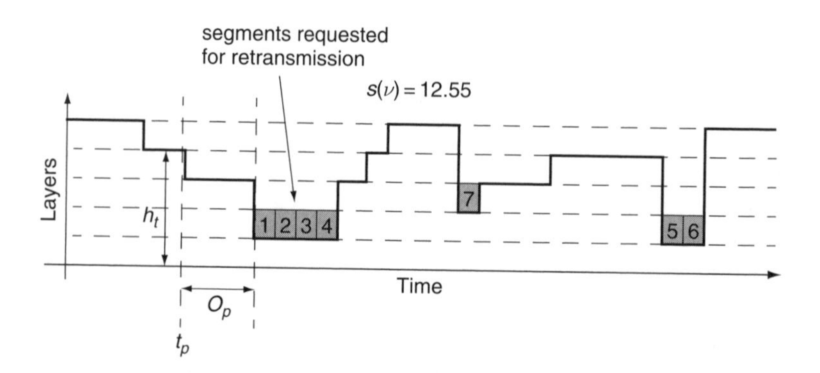

Figure 5.5 U-LLF operation.

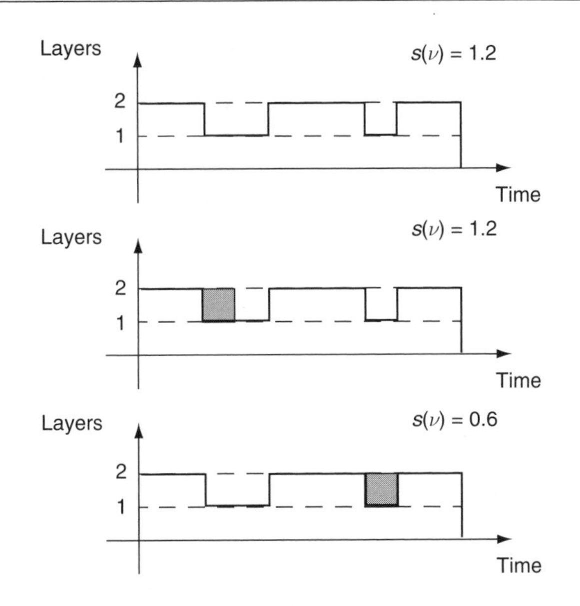

Figure 5.6 Influence of closing gaps on spectrum.

The influence of closing gaps on the spectrum can potentially be quite high. Therefore, in contrast to W-LLF and U-LLF, a prioritization of the missing segments is performed which also takes the closing of gaps into account. This prioritization is achieved by first sorting the segments according to the length of the gap they belong to and then using their layer levels for further sorting. The resulting heuristic is called 'unrestricted shortest gap lowest layer first' (U-SG-LLF). An example of the scheduling of missing segments is given in Figure 5.7.

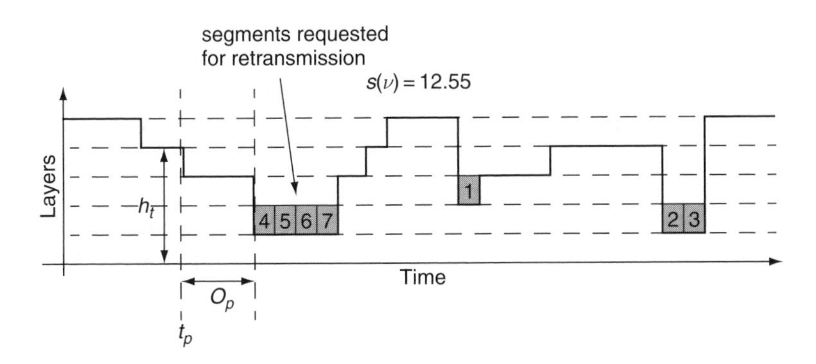

Figure 5.7 U-SG-LLF operation.

5.4.2.3 Unrestricted Lowest Layer Shortest Gap First (U-LL-SGF)

Since it is by no means clear which sorting criterion (i.e., gap length or layer level) should be used first, a further heuristic where missing segments are first sorted by their layer level and then sorted further by gap lengths was investigated. This heuristic is called 'unrestricted lowest layer shortest gap first' (U-LL-SGF). Comparing Figures 5.7 and 5.8 shows that this heuristic can result in a different retransmission schedule.

5.5 SIMULATIONS

In order to compare the different retransmission scheduling algorithms from the previous section and investigate their dependency upon different parameters, a number of experiments based on a custom simulation environment (implemented in C++) were performed.

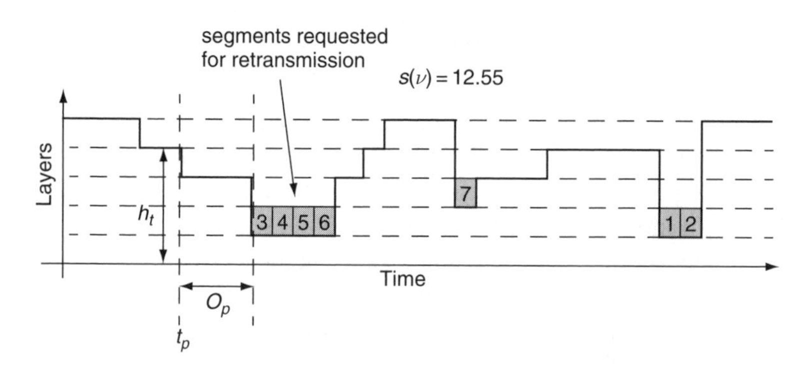

Figure 5.8 U-LL-SGF operation.

The simulations were performed in the following manner. For each simulation an instance of a layered video on the cache is randomly generated. Here, such a layered video instance is modelled as a simple finite birth–death process since it is the result of the congestion-controlled video transmission which restricts state transitions to direct neighbour states. $\{0, \ldots, H\}$ is the state space and birth and death rate are chosen to be equal to $1 - 1/\sqrt{3}$ (for all states) which results in a mean length of 3 time units for periods with stable quality level.† A discrete simulation time is used where one unit of

† The parameter choice is rather arbitrary. However, simulations with other values showed no significant impact on our results.

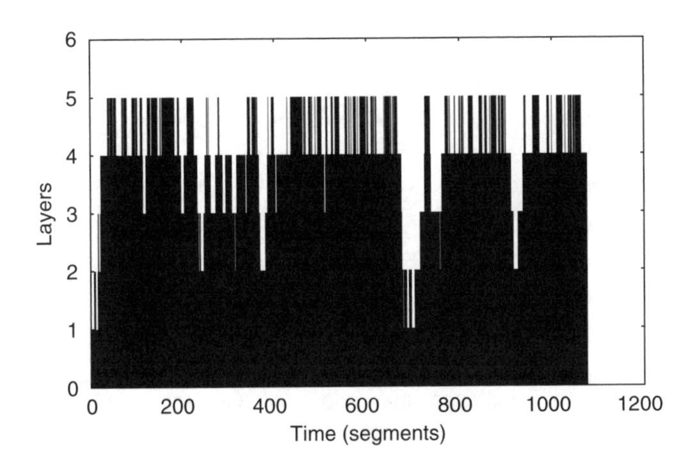

Figure 5.9 Randomly generated layered video on the cache.

time corresponds to the transmission time of a single segment. In Figure 5.9, an example video instance generated in this way is given.

The simulation environment allows the application of the different algorithms described in section 5.3 and the variation of parameters such as the bandwidth available for retransmissions between server and cache. During the simulations, the spectrum (as defined in Chapter 4) of the cached video instances is continuously calculated, and the different algorithms' performance is compared, given parameters such as the available bandwidth. In all simulations a prefetching offset of $O_p = 5$ segments is assumed.

5.5.1 Comparison of the Client-centric Heuristics

At first, a series of 1000 simulations with all retransmission scheduling algorithms from section 5.3 where all parameters were chosen identical (except the window sizes for W-LLF) was performed. This large sample ensured that the 95% confidence interval lengths for the spectrum values were less than 0.5% of the absolute spectrum values for all heuristics. The results for the evolution of the spectrum values for the different algorithms are shown in Figure 5.10.

These results indicate that there is a significant gain with respect to the spectrum of the cached video for the unrestricted retransmission scheduling algorithms in comparison to W-LLF. Of course, if window sizes are chosen large enough for W-LLF, it improves and finally approaches U-LLF.

By taking a closer look at Figure 5.10 it becomes clear that the final spectrum of the U-SG-LLF heuristic is lower by a value of 100 compared to U-LLF and U-LL-SGF. Taking into account the results presented in Table 4.3

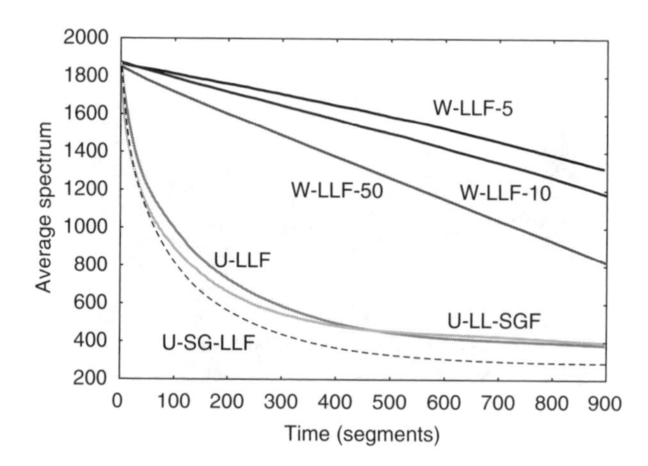

Figure 5.10 Average spectrum of 1000 simulation runs for each heuristic (10 layers, retransmission bandwidth = 2, window size = 5 (W-LLF-5), window size = 10 (W-LLF-10).

where a spectrum improvement by the value of 2 already leads to an increase in the subjective quality, U-SG-LLF can be seen as significantly better than the other heuristics.

In Figures 5.11 and 5.12 the results of one simulation for all four heuristics are shown. The single figures show the differences between the heuristics in more detail. For the case of W-LLF-5 the amount of layer variation is not significantly reduced which, in contrast, is the case for the unrestricted heuristics. Comparing the unrestricted heuristics with each other shows that U-LLF and U-LL-SGF behave almost identically, while the result of U-SG-LLF is quite different and better with respect to the spectrum. All three have in common that a large number of retransmitted segments are located in the second half of the sequence (time segments 500–1000). This effect will be discussed in more detail in section 5.5.3. Altogether, these results reflect what is also shown in Figure 5.10: the spectrum is reduced only by a small amount in the case of W-LLF; U-LLF and U-LL-SGF behave similarly; and U-SG-LLF provides the largest spectrum reduction.

5.5.2 Parameter Dependency Analysis for Client-centric Heuristics

In the following, the heuristics' dependencies on certain parameters are investigated. For all of these simulations, only the U-SG-LLF heuristic is used, since it showed the best performance of all heuristics in the experiment of the preceding section.

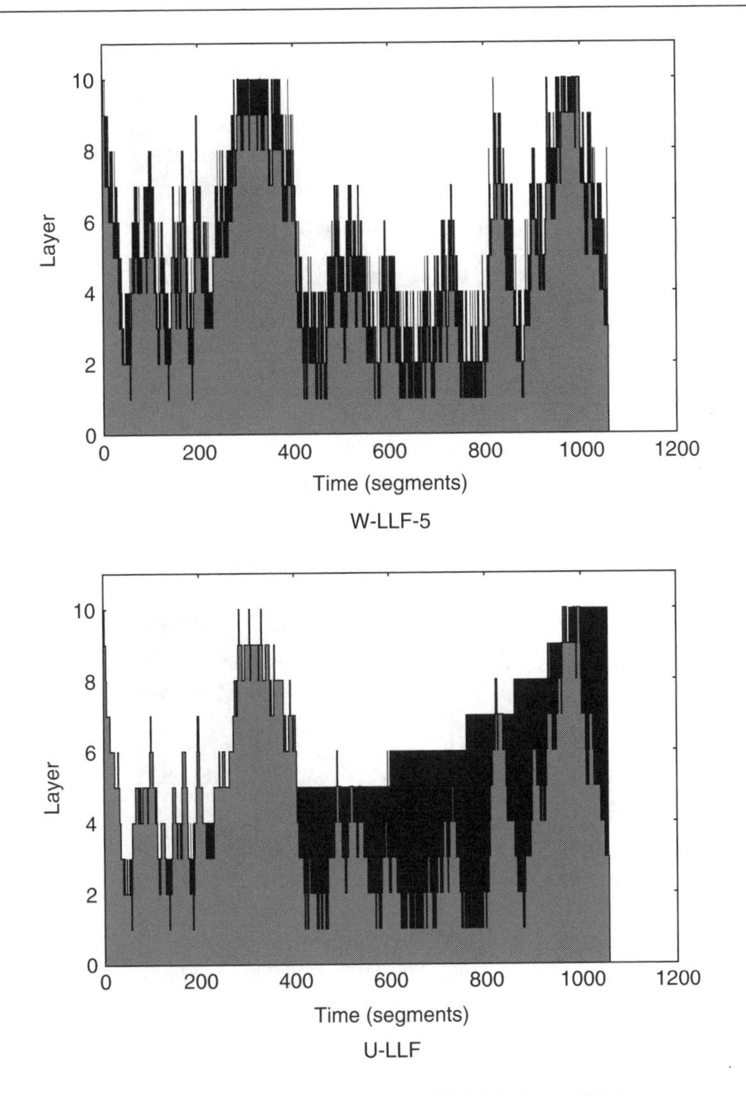

Figure 5.11 Single simulation of W-LLF-5 and U-LLF.

5.5.2.1 Number of Layers (Increasing Bandwidth)

For this simulation, the number of layers per cached video were set to be either 5, 10 or 20 layers. It is assumed that the single layer size is constant and, thus, the bandwidth needed to stream a 20-layer video is twice as much as that needed for a 10-layer video. To isolate the effect of encoding videos with different numbers of layers, the available retransmission bandwidth was scaled in proportion to the number of layers, i.e., for 5 layers 2 segments

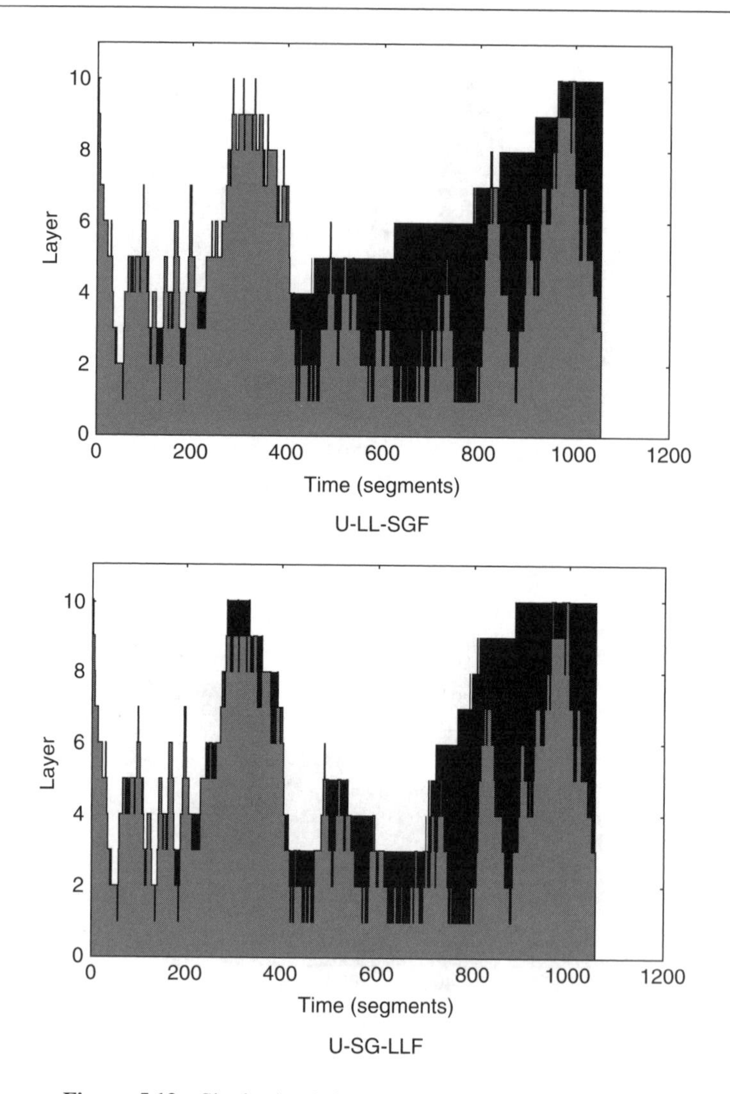

Figure 5.12 Single simulation of U-LL-SGF and U-SG-LLF.

of retransmission bandwidth per time unit were assumed, for 10 layers 4 segments, and for 20 layers 8 segments. For each of these three alternatives 1000 simulations were performed and again the average of the spectra over time was calculated.

As Figure 5.13 shows, the spectrum converges for both the 5- and 10-layer video. For the 20-layer video the spectrum increases slightly in the end. This effect is explained in section 5.5.3. Yet, the higher the number of

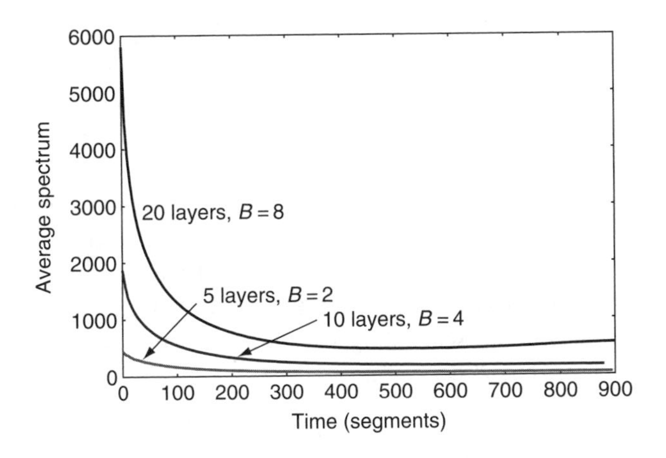

Figure 5.13 Different number of layers.

layers, the higher the average spectrum. This is intuitive because the more fine-grained the layered encoding, the more variations may be introduced during a congestion-controlled transmission and the harder it is for the retransmission scheduling to smooth these variations. This result has implications for the use of FGS as encoding format. Although FGS basically consists of two layers, depending on the actual transmission into the cache, the number of different levels in layer two could be quite high. Thus, retransmission scheduling for FGS would perform equally with or worse than the presented 20-layer case.

5.5.2.2 Number of Layers (Constant Bandwidth)

For this simulation, the number of layers per cached video are determined to be either 5, 10 or 20. Increasing the number of layers in this case does not increase the maximum bandwidth of the layer-encoded video. Instead, the bandwidth of each single layer is decreased. In our work on *fair share claiming* ([92] and Chapter 7), we investigate how a TCP-friendly transmission exploits its fair share of network resources, taking into account that the constrained granularity of layer-encoded video inhibits an exact adaptation to actual transmission rates.

The result of this investigation shows that available bandwidth for retransmission is reduced with an increasing number of overall layers. Thus the bandwidth available for retransmission was not increased as in the simulation presented above. The retransmission bandwidth is set to one segment per time unit for all three types of video objects. Here too 1000 simulations were run for each of the three heuristics. As Figure 5.14 shows, the spectrum converges for each of the three alternatives. A comparison to the results presented in Figure 5.13 shows that the final average spectrum for the 10- and 20-layer cases is higher because the number of segments that could be retransmitted is lower owing to a reduced relative bandwidth. This simulation shows that one retransmission phase may not always be sufficient to obtain full quality for the cached content, since the spectrum is not reduced to 0. Here the drawback of FGS becomes even more obvious, if we take into account the remarks made in the section above. With the expected high amount of layer variation, retransmission scheduling will not perform as well with FGS as with discrete layer-encoding formats.

5.5.2.3 *Available Retransmission Bandwidth*

In the next set of simulations, the effect of different amounts of available retransmission bandwidth on the performance of U-SG-LLF was investigated.

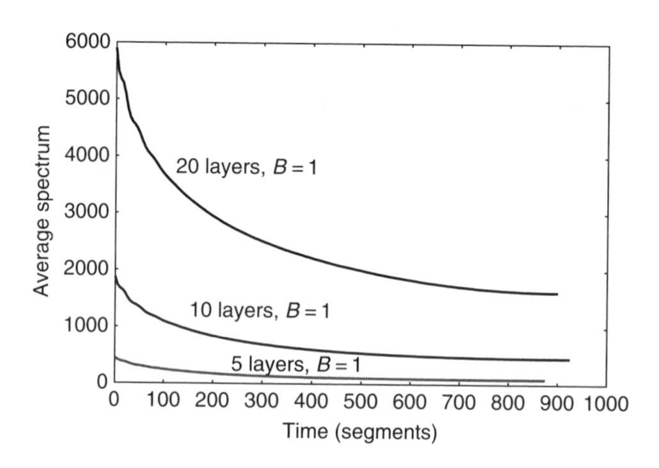

Figure 5.14 Different number of layers (constant bandwidth).

Not surprisingly, the spectrum converges more quickly with a higher available retransmission bandwidth, as shown in Figure 5.15. The reason for the very similar spectrum curves for $B = 6$ and $B = 10$ is that there is sufficient retransmission bandwidth for both cases, which makes it possible to retransmit all missing segments that have a playout time greater than 500. Owing to the prefetching offset, missing segments from the beginning cannot be retransmitted and, therefore, a spectrum of 0 is not achieved.

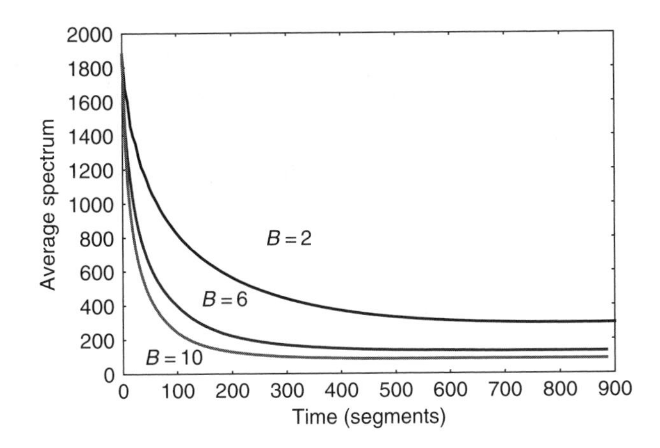

Figure 5.15 Different amounts of available retransmission bandwidth.

5.5.2.4 Amount of Layer Variation

Depending on the situation on the network the amount of layer variation that is introduced by the adaptive transmission can vary. To investigate the behaviour of the amount of layer variation on the retransmission scheduling heuristics, an additional investigation was performed in which the amount of that variation was different. Three different simulations were performed assuming that the number of layer variations was 100, 200 and 400, respectively. As shown in Figure 5.16(a), the value of the spectrum is proportional to the amount of layer variation. Observing the relative reduction of the spectrum through the algorithm shows (see Figure 5.16(b)) that it performs almost identically, independently of the initial amount of layer variation. The relative spectrum is obtained as the ratio between the actual value of the spectrum and the maximum value for each time step.

5.5.2.5 Initial Transmission Quality

Finally, a series of simulations were performed in which different initial transmission qualities were assumed, resulting in cached videos where the maximum number of cached layers is lower than the maximum number of layers for the original video. In contrast to the preceding experiments, the spectrum values are not sampled but a single simulation is used since the relevant effects can be shown in more detail. For each simulation, the ratio between the maximum of cached layers (MCL) and the maximum of original layers (MOL) is modified.

As Figure 5.17 shows, spectrum values start to rise again for the last third of the video. This effect is especially pronounced for worse initial transmission qualities (MCL/MOL = 10/17). To intensify the illustration of this effect the bandwidth available for retransmissions is proportionally increased according to the MOL and is set to 2, 4 and 8, respectively.

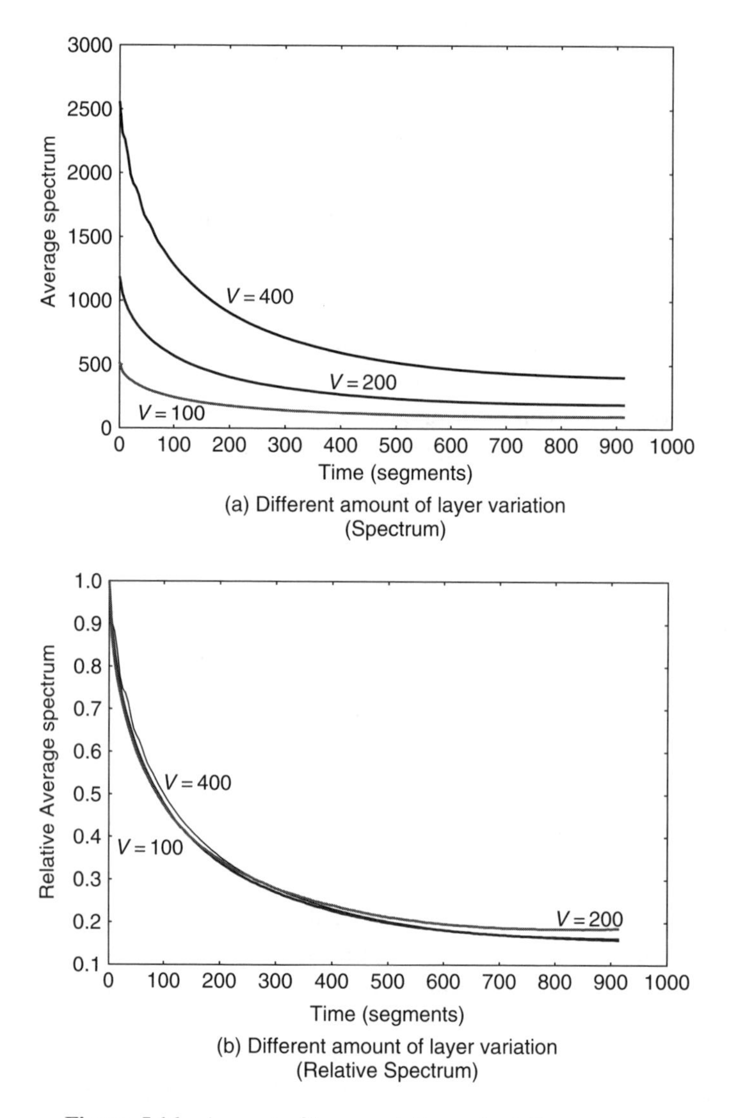

Figure 5.16 Amount of layer variation and relative spectrum.

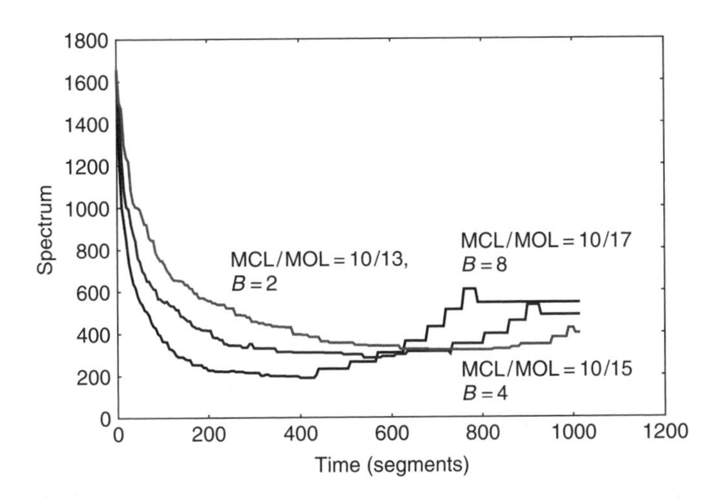

Figure 5.17 Influence of initial transmission quality.

Looking at the cached video that results from the retransmission scheduling heuristic (U-SG-LLF) in Figure 5.18 sheds light on the reason for this effect. It can be observed that the retransmission scheduling 'builds a staircase' at the end of the cached video which is not beneficial with respect to the minimization of the spectrum. The reason for this behaviour is that the algorithm only considers missing segments ahead of the playout time $(t_p + O_p)$. Thus, if all gaps are closed the algorithm starts to request segments from the next layer starting from $t_p + O_p$. This happens several times, leading

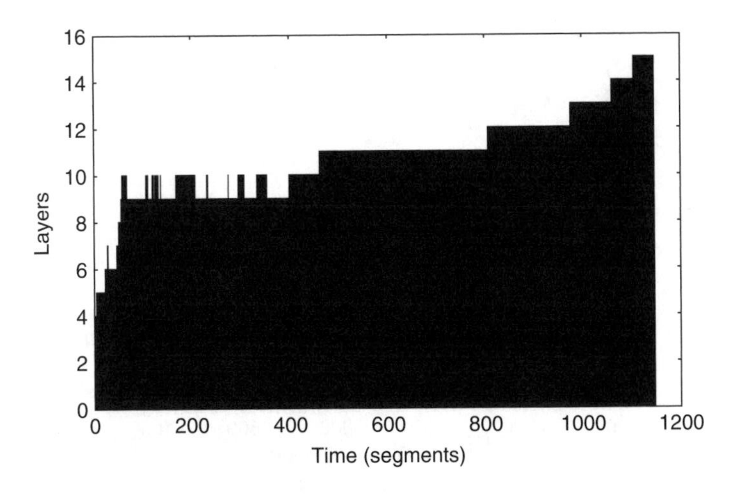

Figure 5.18 Cached video after retransmission phase.

to the staircase shape exhibited in Figure 5.18. A retransmission scheduling based technique will be presented in the following section, while Chapter 6 introduces a more general approach technique called polishing.

5.5.3 Totally Unrestricted Heuristics

As a consequence of the results from section 5.5.2, a further investigation should show how the modification of the heuristics in a way that retransmissions are not limited to segments that are located after $t_p + O_p$, could cure the problem the heuristics had with bad initial transmission qualities. However, it has to be observed that such a totally unrestricted retransmission scheduling algorithm brings the possibility that retransmitted segments may arrive too late for the current client and might thus be retransmitted in vain if no other client ever requests that video. Thus, some of the attractiveness of write-through caching is lost. On the other hand, it also offers the chance to obtain a complete copy of the video on the cache.

The simulations from section 5.5.2 were repeated with the now totally unrestricted version of U-SG-LLF. The results are shown in Figure 5.19. Obviously, the problem of rising spectrum values is solved. This observation is reinforced when the cached video as it results from the totally unrestricted heuristic in Figure 5.20 is compared to its counterpart in Figure 5.18.

However, in order to assess how many segments would be scheduled for retransmission which could no longer be viewed by the current viewer, these

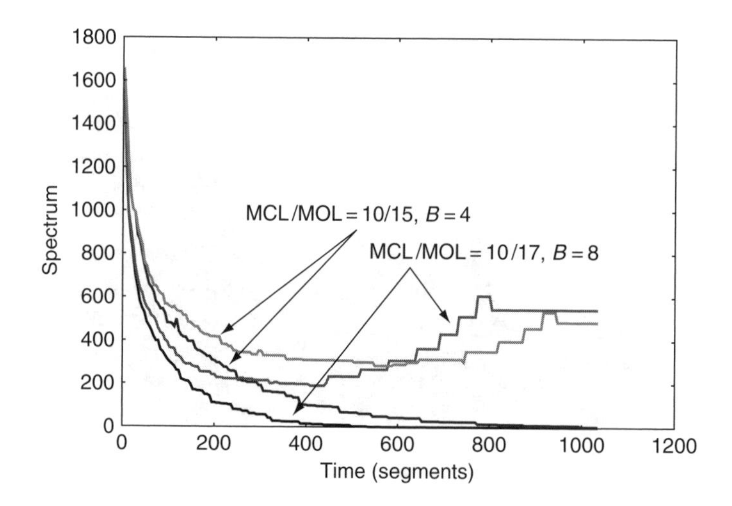

Figure 5.19 Comparison of restricted and totally unrestricted heuristics.

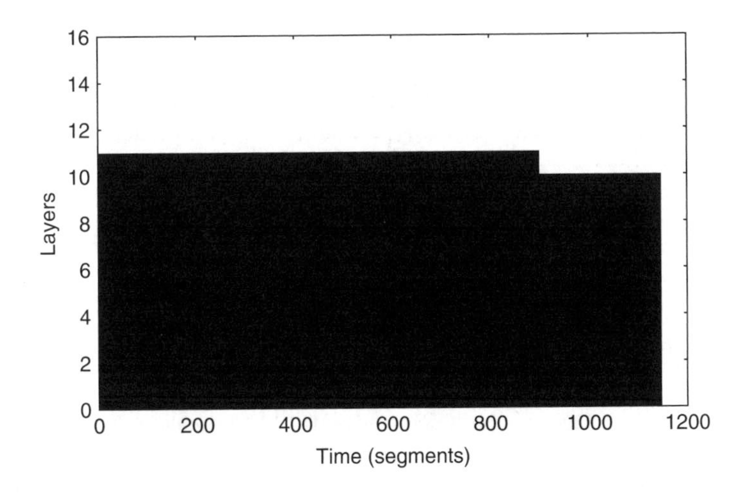

Figure 5.20 Cached video after retransmission phase for totally unrestricted heuristic.

'late' segments were also recorded: with MCL/MOL = 10/15 38% and with MCL/MOL = 10/17 37% of the retransmitted segments arrived too late. This is certainly a substantial number of late segments and, thus, one has to make a decision between generating a fairly smooth cached video and using all available retransmission bandwidth to benefit the current client.

5.6 CACHE-CENTRIC RETRANSMISSION SCHEDULING

The heuristics presented in section 5.4 are designed to maximize the perceived quality for the current viewer of the video. Since this can have a negative effect on the cached video object, as shown in Figure 5.19, and previous results from section 5.5.3 imply a possible solution for this problem, the unrestricted heuristics presented in section 5.4 are modified. With this modification segments for which the playout time, t_p, has already passed, can also be scheduled. Thus, the overall quality of the cached video is improved for all potential viewers and not only the current one. This rather *social* approach is called *cache-centric* retransmission scheduling in contrast to *viewer-centric* retransmission scheduling. Since it is debatable which of the strategies (client- or cache-centric) should be applied for retransmission scheduling, a possible cache operator should have the chance to choose between both. To investigate the behaviour of the heuristics for cache-centric retransmission scheduling, an additional set of simulations was performed.

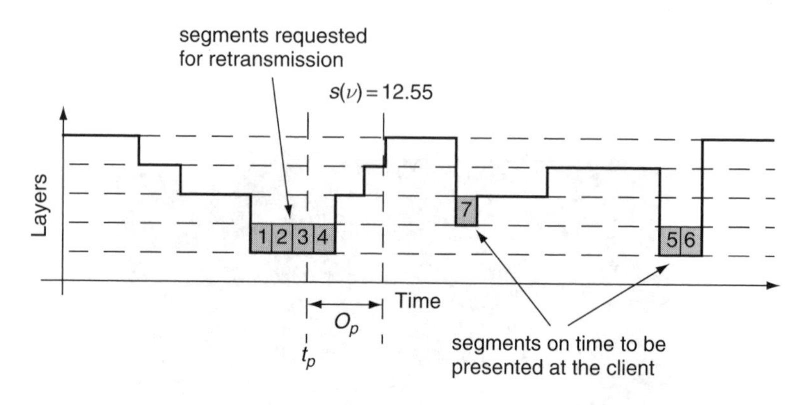

Figure 5.21 Cache-centric U-LLF operation.

As a representative for all unrestricted heuristics Figure 5.21 depicts the behaviour of the cache-centric U-LLF heuristic.

The simulation environment is identical to the one presented in section 5.3 with the exception that the implementation for the unrestricted heuristics was slightly modified as described above.

5.6.1 Comparison of the Cache-centric Heuristics

As in the simulation for the viewer-centric retransmission scheduling heuristics, a series of 1000 simulations with all three modified retransmission scheduling algorithms from section 5.6, where all parameters were chosen to be identical, was performed. This investigation was performed in order to see whether the cache-centric retransmission scheduling heuristics would behave in a different way. As can be seen in Figure 5.22 the results of the simulations meet the expectations. The final spectrum of the cache-centric heuristics is significantly lower than the spectrum of the viewer-centric ones. (To show this difference in more detail the result for U-SG-LLF from section 5.5.1 is shown in Figure 5.22, too.) The final spectrum of the cache-centric U-SG-LLF heuristics is lower by a value of 217 compared to the viewer-centric U-SG-LLF heuristic.

As with the simulations for the viewer-centric heuristics, of all the algorithms, U-SG-LLF performed best. Also here the results of a single simulation for all three heuristics are shown. As can be seen from Figure 5.23 the staircase-effect introduced by the heuristics from section 5.4 does not occur with the totally unrestricted heuristics. Similarly to the results presented in Figure 5.11, U-LLF and U-LL-SGF perform almost identically while the outcome of U-SG-LLF is different.

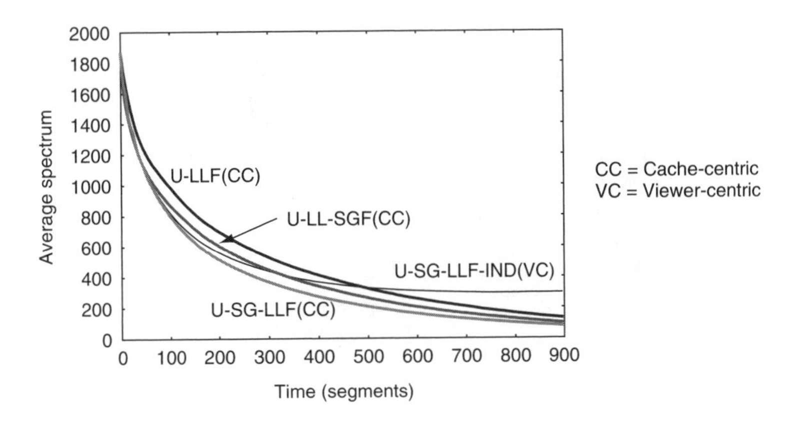

Figure 5.22 Average spectrum of 1000 simulation runs for each heuristic (10 layers, retransmission bandwidth = 2).

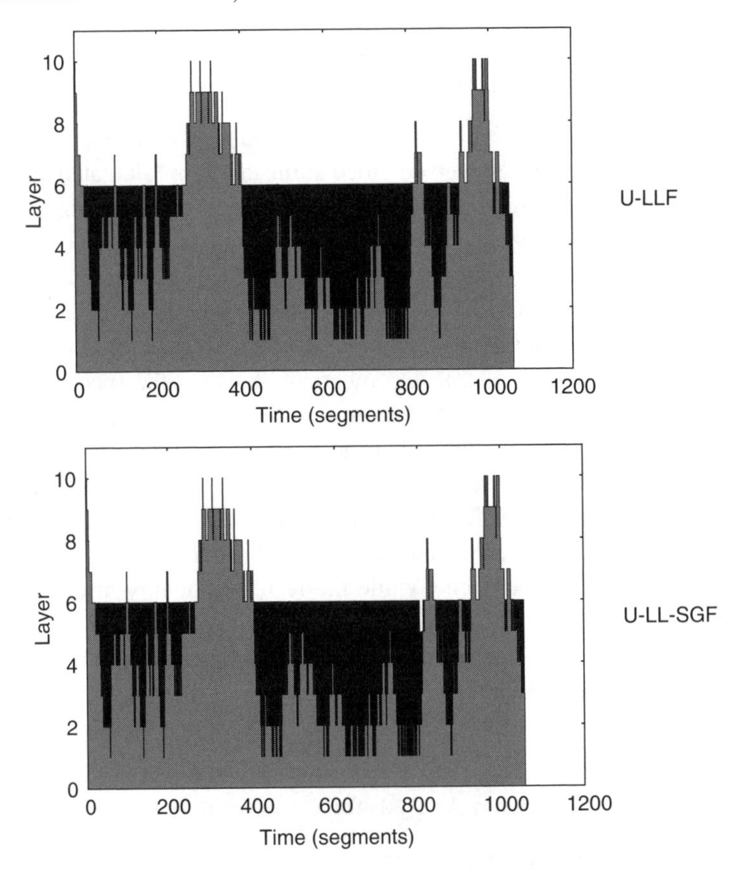

Figure 5.23 Single simulation of totally unrestricted heuristics.

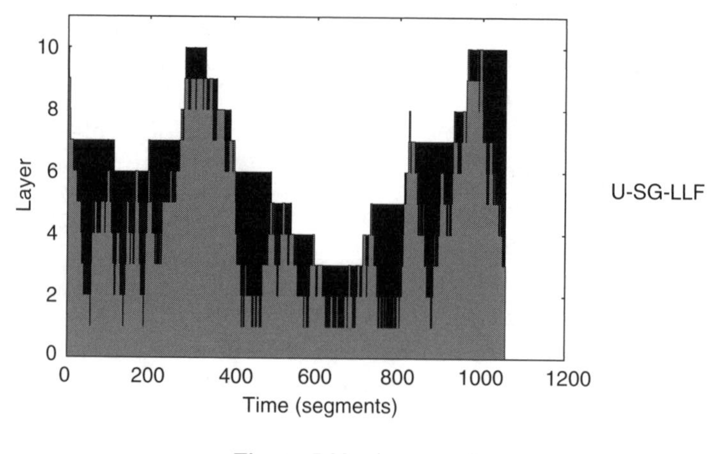

Figure 5.23 (continued)

5.6.2 Spectrum for the Current Viewer

To assure that the cache-centric approach is also beneficial for the current viewer the spectrum of the received video at the client is calculated additionally. As can be seen in Figure 5.24(a) the spectrum at the client is significantly higher than that at the cache. The resulting spectrum at the cache is almost reduced to a value of 0, while the resulting spectrum at the client is a least 200 (for the case of U-SG-LLF). The cause for the increased spectrum at the client is based on the timing constraints for the playout of the single segments of the video (see section 5.5.3). A comparison between the spectrum of the viewer-centric approach and the spectrum at the client for the cache-centric approach (Figure 5.24(b)) shows that the latter is only slightly worse. The spectrum at the cache is equal to the spectrum at the client in the case of the viewer-centric approach and, thus, is worse than that for the cache-centric approach. Therefore, the cache-centric approach has the advantage of achieving a better quality at the cache, while the resulting quality at the client is only slightly worse compared to the viewer-centric approach. In section 5.7, an approach is presented which combines the benefits of the client-centric and cache-centric approaches.

5.6.3 Parameter Dependency Analysis for Cache-centric Heuristics

A parameter dependency analysis is also performed for the cache-centric heuristics. Similarly to the analysis for the viewer-centric heuristics only the

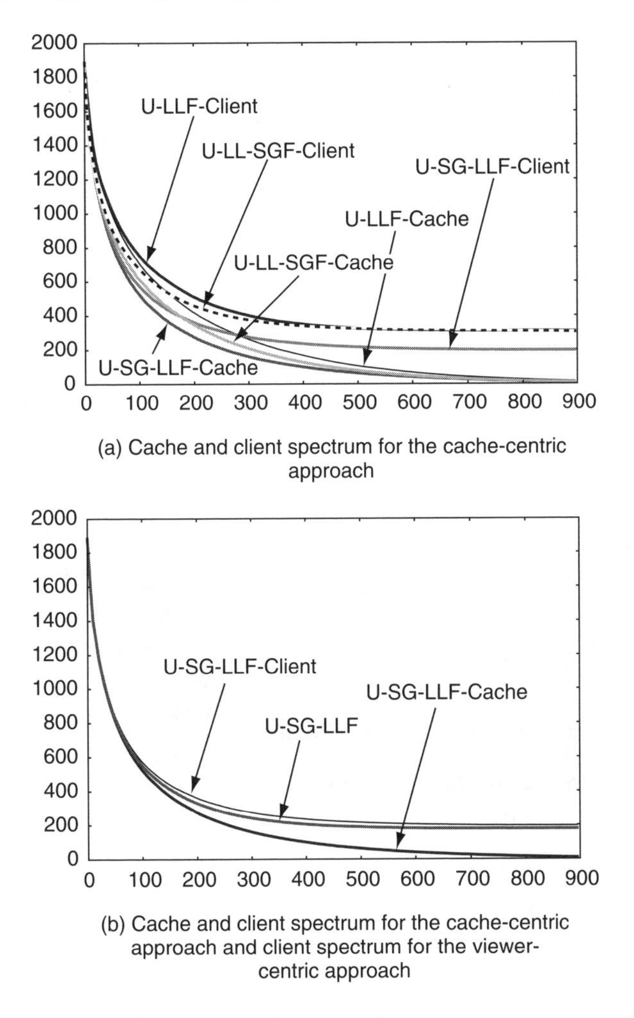

Figure 5.24 Cache vs. client spectrum.

U-SG-LLF heuristic is looked at for this analysis, since it showed the best performance of all heuristics in the experiment of the preceding section.

5.6.3.1 Number of Layers (Increasing Bandwidth)

As in the simulation in section 5.5.2 the number of layers per cached video was set to be either 5, 10 or 20 layers. In this simulation it is assumed that the single layer size is constant and, thus, the bandwidth needed to stream a 20-layer video is twice as much as the one needed for a 10-layer video. The

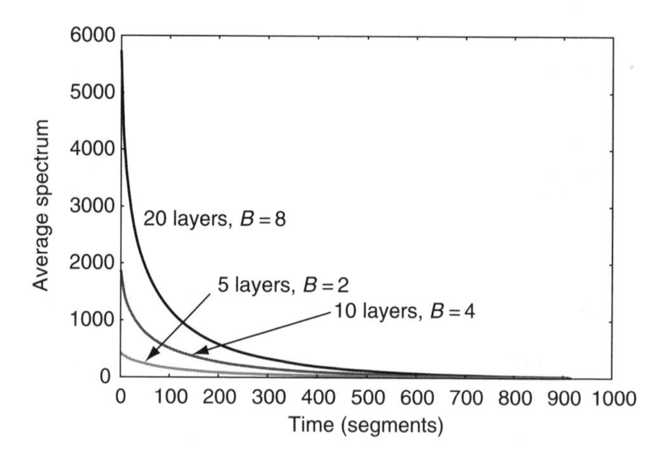

Figure 5.25 Different number of layers (increasing bandwidth).

results of the simulation (shown in Figure 5.25) are similar to those for the viewer-centric heuristic, with the difference that the final spectrum reaches a lower value.

5.6.3.2 Number of Layers (Constant Bandwidth)

Also here the results of the simulation are similar to the corresponding ones in section 5.5.2. Note that the values for the final spectra (Figure 5.26) are

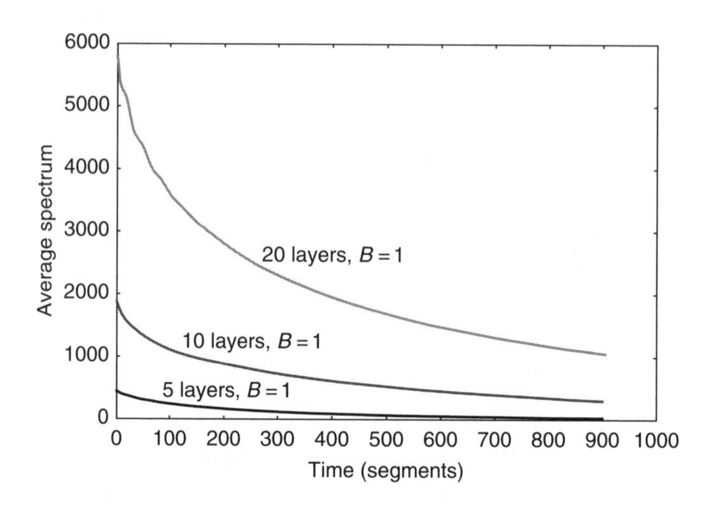

Figure 5.26 Different number of layers.

lower than in the case of the viewer-centric heuristic (Figure 5.15). As in section 5.5.2, one retransmission run might not be sufficient to retransmit all missing segments to the cache. (Even for the object that consists of up to 5 layers the final spectrum is larger than zero.)

5.6.3.3 Available Retransmission Bandwidth

In the next set of simulations, the effect of different amounts of available retransmission bandwidth on the performance of cache-centric U-SG-LLF was investigated. Not surprisingly, the spectrum converges more quickly with a higher available retransmission bandwidth, as shown in Figure 5.27. The reason for the very similar spectrum curves for $B = 6$ and $B = 10$ is that there is sufficient retransmission bandwidth for both cases, which allows one to retransmit all missing parts of the cached video. Thus, in contrast to the simulation in section 5.5.2 the video is stored on the cache in full quality.

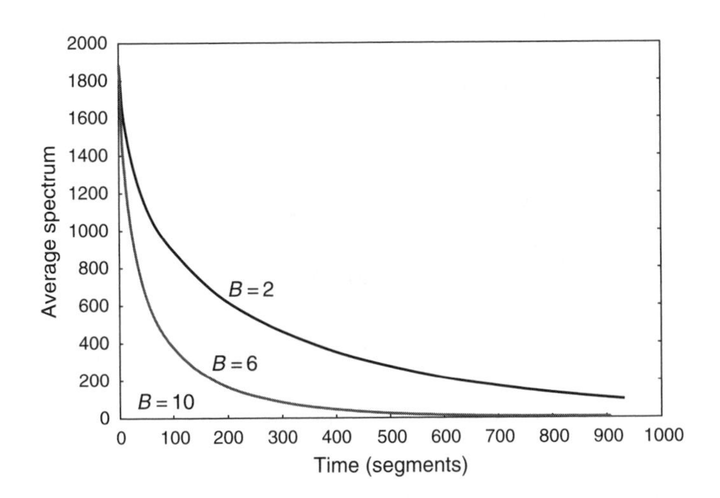

Figure 5.27 Different amounts of available retransmission bandwidth (10 layers).

5.6.3.4 Amount of Layer Variation

The result of the simulation on the amount of layer variation on the cache-centric U-SG-LLF heuristic is shown in Figure 5.28. As expected, here too the spectrum is proportional to layer variations but the convergence is stronger for the cache-centric heuristic, as can be seen from the relative average spectrum of this simulation.

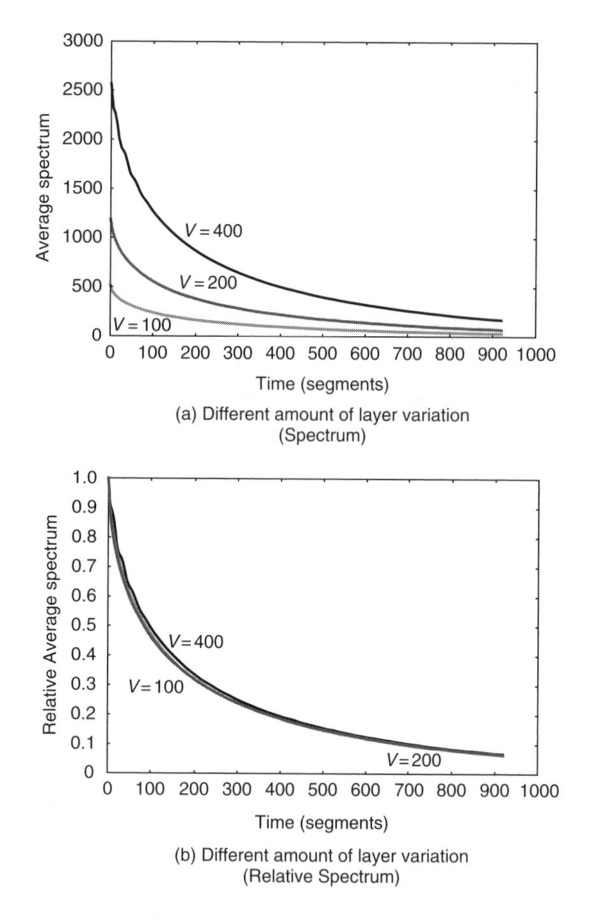

Figure 5.28 Amount of layer variation.

5.7 CACHE-FRIENDLY VIEWER-CENTRIC RETRANSMISSION SCHEDULING

One special case that can occur with viewer-centric retransmission scheduling is the one in which no more segments are retransmitted, although there are still missing segments. This is the case when all missing segments with a playout time greater than t_p are retransmitted, as shown in Figure 5.29 for the U-LLF heuristic. At this point in time there exist no further segments which have a playout time greater than t_{p}, but there is still the possibility to transmit missing segments with a playout time that has already passed.

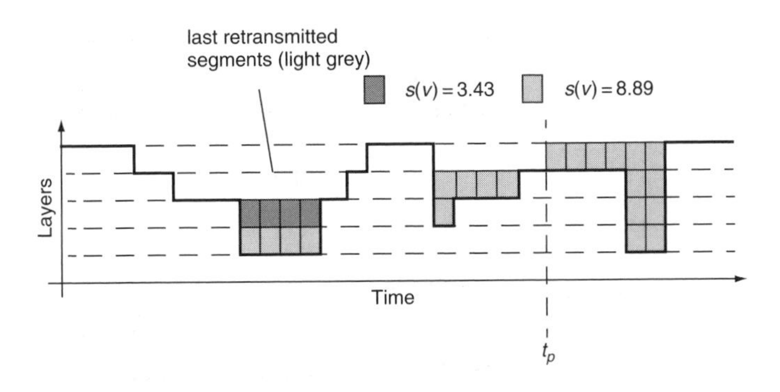

Figure 5.29 Cache-friendly viewer-centric U-LLF operation.

Those segments are useless at the client and can only be used to improve the quality of the cached content. Thus, they should not be forwarded from the cache to the client. This approach is called *cache-friendly viewer-centric* retransmission scheduling.

In Figure 5.29, only the dark grey segments are forwarded to the client since they arrived at the cache early enough to allow a timely playout at the client. The light grey segments are useless at the client but improve the quality of the cached video which can be beneficial for other clients requesting that video at a later point in time. The perceived quality for the viewer is exactly the same as in the case of plain viewer-centric retransmission scheduling, while the resulting quality of the cached video is always better in the case of cache-friendly viewer-centric retransmission scheduling. As shown in Chapter 7, the additional bandwidth that is needed to transmit those cache-friendly segments does not influence the quality of the stream that is forwarded to the client, if the fair share claiming transport mechanism is used. In this case, missing segments are only retransmitted if enough additional bandwidth on the link between server and cache is available.

5.7.1 Simulations

Using the same simulation environment as described in section 5.5, a series of simulations were performed with the cache-friendly viewer-centric heuristic. In Figure 5.30(a) the different retransmission scheduling approaches (cache-centric, viewer-centric, cache-friendly viewer-centric) for the U-SG-LLF heuristic are compared with each other. The comparison of the three different heuristics for the cache-friendly viewer-centric approach (Figure 5.30(b)) confirms the results of the simulations presented in sections 5.5.1 and 5.6. Also in the simulation for the approach presented here U-SG-LLF is the

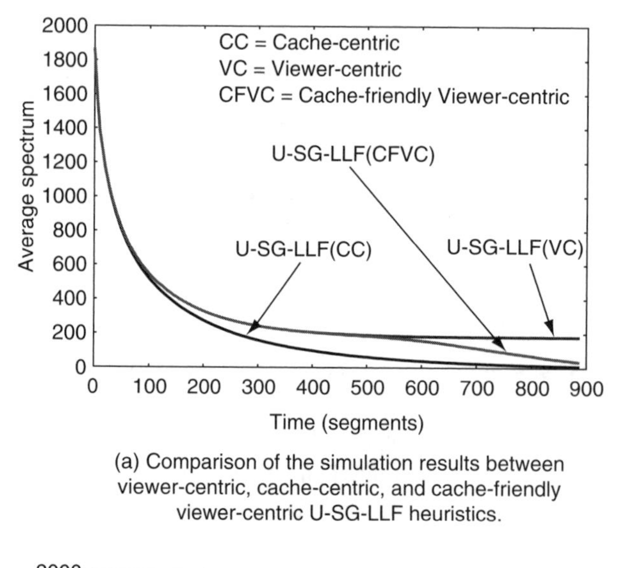

(a) Comparison of the simulation results between viewer-centric, cache-centric, and cache-friendly viewer-centric U-SG-LLF heuristics.

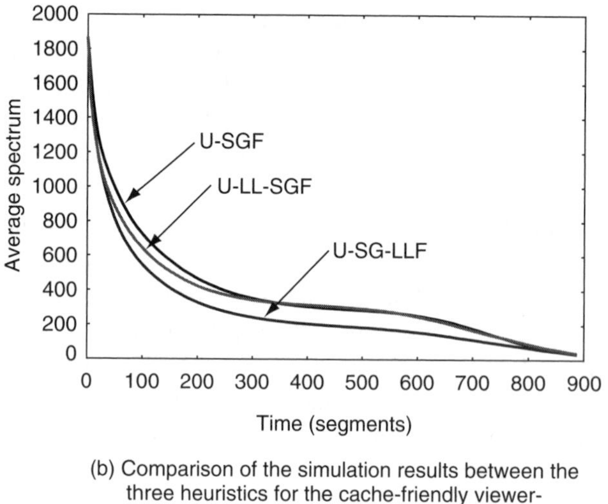

(b) Comparison of the simulation results between the three heuristics for the cache-friendly viewer-centric approach.

Figure 5.30 Average spectrum of 1000 simulation runs for each heuristic (10 layers, retransmission bandwidth = 4).

best performing heuristic, while U-LLF and U-LL-SGF are similar in their performance. Since U-SG-LLF again is the best performing of the three heuristics, it was chosen for the comparison between the different approaches depending on the retransmission focus. As shown in Figure 5.30, the cache-friendly viewer-centric approach results in a better spectrum on the cache than the plain viewer-centric approach. On the other hand, the resulting spectrum for the cache-friendly viewer-centric approach is slightly higher than that for the cache-centric approach. Note that the resulting spectrum for the cache-friendly approach at the client is identical to the spectrum of the plain viewer-centric approach.

5.8 SUMMARY

The work presented in this chapter has focused on the problem of how to deal with retransmissions of missing segments for a cached layered video in order to meet viewers' demands to watch high-quality video with relatively little quality variation. After the introduction of the basic retransmission scheduling approaches and the motivation of the scheduling goal, the complexity of the optimal retransmission scheduling algorithm and the drawbacks of this algorithm were shown. Those results led to the development of different retransmission scheduling heuristics. The development of these heuristics was influenced by the results of the subjective assessment on variations in layer-encoded video (see Chapter 4) and preceding work by Rejaie *et al.* [115]. Based on the retransmission focus, that means, should the current viewer or the cached version of the video be maximized, three different approaches, viewer-centric, client-centric, and cache-friendly client-centric, are proposed.

A custom simulation environment was built to compare the proposed heuristics with each other. This simulation not only allowed a comparison of the single heuristics between each other but, additionally, the performance of the heuristic when used in the scope of a certain retransmission approach (viewer-centric, cache-centric, or cache-friendly viewer-centric) could be compared. In addition, simulations on the dependency of the heuristics on system parameters were performed.

The results of the simulations show that the newly created heuristics can outperform an existing one. They also show that one of the three new heuristics (U-SG-LLF) outperforms the remaining two. The comparison of the results for the different retransmission approaches shows that the cache-friendly viewer-centric approach can be seen as a good compromise, since the quality for the current viewer is maximized, while the quality of the cached object can be increased to close to the maximum.

Nevertheless, administrators of a VoD service that makes use of retransmission have the freedom to choose one of the three approaches based on their preferences without requiring significant changes on the overall SAS architecture.

6

Polishing

6.1 MOTIVATION

The combination of caching and adaptive streaming bears the disadvantage that a layer-encoded video can only be cached according to the available bandwidth of the path between server and cache, if a congestion-controlled transport mechanism is applied. The created copy of the video stored on the cache might, depending on the path's condition, contain a large amount of layer variations. Thus, forwarding such a video object towards a client might be annoying for a viewer.

In this chapter, a new technique, called polishing, is presented. With polishing a cache considers sending only a subset of the segments of a locally stored object in order to reduce layer variations to the client. Investigations by performing a subjective assessment of layer variations in videos (see Chapter 4) have shown that it can be beneficial to omit the transmission of certain segments, especially if the amount of layer variation is thereby reduced.

At first, this might sound rather strange since some information is not transmitted at all and, thus, the PSNR of the video is reduced. Yet, investigations in Chapter 4 have shown that, despite reducing the PSNR, reducing layer variations can increase the perceived quality of a video, if done carefully. That is the case, in the previously performed subjective assessment, if a layered video with a lower PSNR but less layer variation was given a higher perceptual quality than the same video with a higher PSNR and a greater amount of layer variation.

Reconsidering the example from section 2.6, polishing can be applied in two cases. In the first case, a connection to the server that stores the original version of the video object cannot be established but a cached version of the

Scalable Video on Demand: Adaptive Internet-based Distribution M. Zink

video object already exists on the local cache. Assuming that not all missing segments have been retransmitted so far, polishing can be applied to reduce quality variations. In the second case, the storage space at the local cache is exhausted and new video objects should be stored on the cache. Instead of completely removing less popular objects from the cache, polishing can be applied to remove a certain number of segments. Thus, new objects can be added to the cache's storage while only segments of certain objects are removed instead of complete objects.

In the remainder of this chapter an optimal polishing technique is specified and compared with a less complex heuristic by performing a series of simulations. In additional simulations the effect of using polishing for fine-grained cache replacement strategies is investigated.

6.2 POLISHING AND ITS APPLICATIONS

In this section, the fundamental ideas behind polishing are presented. It is shown how polishing can be applied to increase the quality when an incomplete video object is transmitted from the cache to the client. In addition, a cache replacement method based on polishing is presented.

6.2.1 Transport

As mentioned above, the amount of variation in a layer-encoded video that is stored on the cache can be reduced by omitting the transmission of certain segments from the cache to the client.

The challenge of polishing is to identify segments that should not be transmitted in order to increase the perceived quality of a video object at a client. Polishing is a new technique that determines those segments that should not be transmitted from a cache to a client with the goal of increasing the perceived quality of a video at a client. After the caching process of a certain video object is finished, the polishing algorithm (as described in section 6.4) is executed. This algorithm identifies segments that should not be played out in subsequent streaming sessions from cache to clients. Note that the identified segments are not removed from the cache. If this video object is requested, the transport mechanism of the cache will take into account the information gained by polishing and decide which data will be sent to the client and which not.

The information gained by polishing can, for example, be used to stream a polished version of a video object if retransmission scheduling cannot be performed. This might be the case if the server or the link between the server and the cache is down or the server does not have additional capacity to allow retransmission in addition to already active streams. The general concept of polishing in such a case is shown in Figure 6.1.

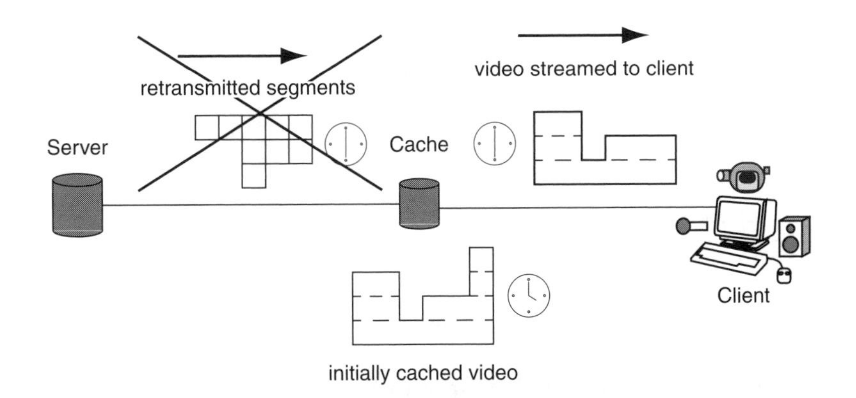

Figure 6.1 Polishing in the case of impossible retransmissions.

6.2.2 Fine-grained Cache Replacement

Since polishing identifies segments that are of less importance in relation to quality, the information gained by applying this technique can also be used for cache-replacement strategies. Assuming that the storage space on a cache is exhausted and data has to be removed from the cache in order to allow the caching of a new object, segments identified by polishing can be deleted. A fine-grained replacement scheme based on segments can increase the efficiency of the cache as shown in reference [118]. If, in addition, popularity information is taken into account, for example, segments of the least popular object are deleted first, the quality of the cached content is also based on its popularity. Figure 6.2 shows the basic principle of cache replacement with the aid of polishing. In this example, four objects are stored on the cache. After performing polishing, 15 segments are identified by the algorithm that can be removed from the cache and, thus, new storage space is created to store an additional object on the cache. The second alternative in Figure 6.2 shows the case in which popularity-based caching is applied. The popularity of the cached object decreases from left to right, influencing the number of segments that are identified for removal by polishing, leading to the fact that objects with a higher popularity are cached in a better quality.

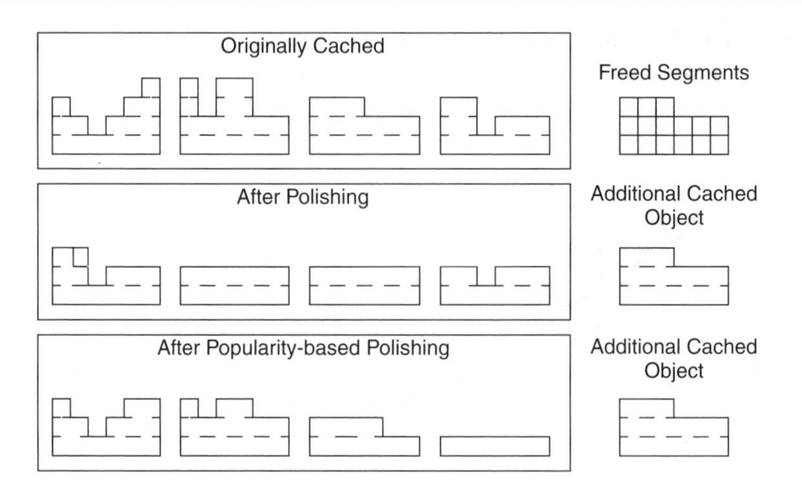

Figure 6.2 Cache replacement with the aid of polishing.

6.2.3 The Spectrum in Combination with Polishing

Although, the spectrum is a good metric for the quality of layer-encoded video (see section 4.6) and, thus, in the case of retransmission scheduling the minimization of the spectrum is the goal to strive for (see section 5.1.3) this goal cannot directly be applied for polishing. The spectrum becomes zero for the case that no layer changes occur, irrespective of how many layers the video object consists of. In the case of retransmission scheduling where new segments are added to the video owing to the fact that any additionally available bandwidth is used for retransmissions, achieving a spectrum of zero always leads to a better quality than that of the originally cached object. In the case of polishing, this assumption is not valid since segments are discarded from instead of added to the layer-encoded video. If the decision to drop certain segments were solely driven by the spectrum, polishing could lead to the fact that all segments of incomplete layers are discarded. This effect, which is denoted as *over-polishing* (see Figure 6.3), is undesirable because it decreases the quality in a drastic manner. In section 6.4, a new algorithm for polishing that avoids the problem of *over-polishing* is presented.

6.2.4 Example

In this section, a simple example to demonstrate the effect of polishing is given. It is assumed that a layer-encoded video is stored in the cache as shown in Figure 6.4. The variations in the number of layers are caused by a congestion controlled transmission between server and cache which results from the network conditions on the path between both (see section 6.5 for details on how the congestion-controlled transmission was simulated). Figure 6.4 also shows the layer-encoded video as it would be transmitted to the client after the polishing algorithm has been performed. In addition, Figure 6.4 shows the result of a simple heuristic where only the highest (5th) layer is dropped. Further details about the optimal polishing algorithm are given in section 6.4. Figure 6.4 shows a significant reduction in layer variations

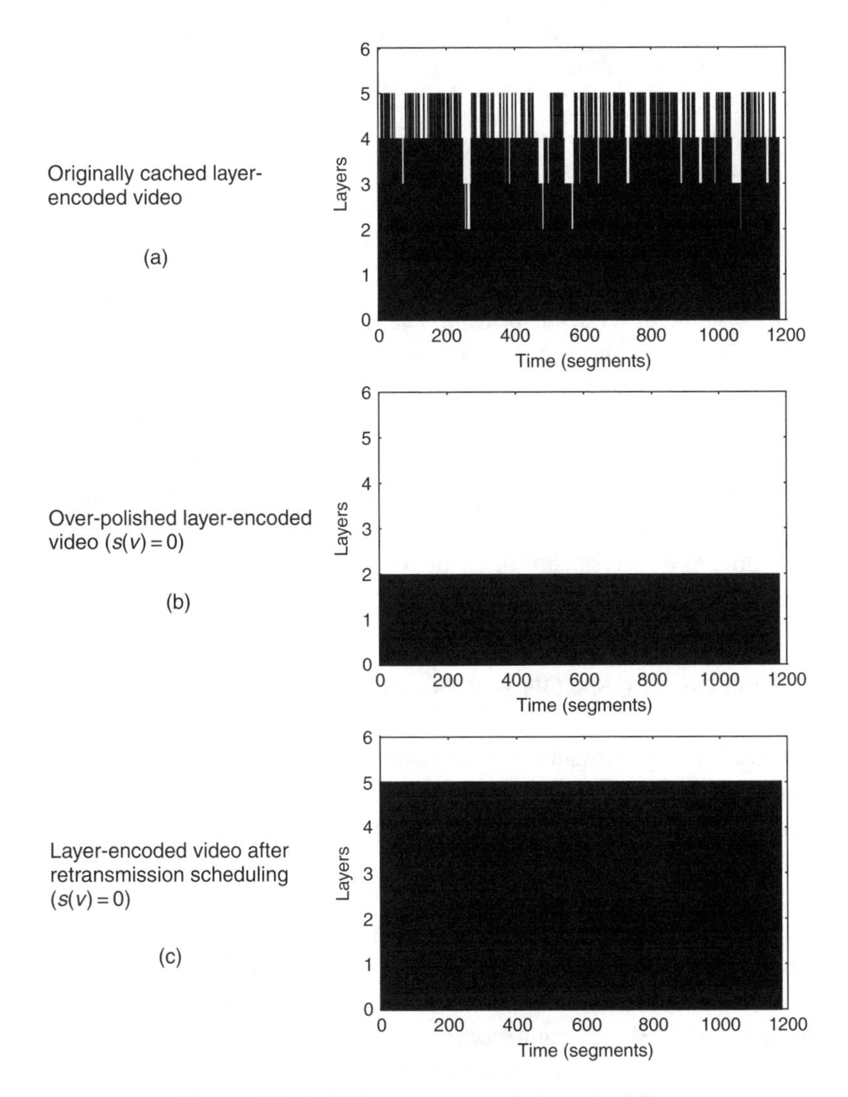

Figure 6.3 Polishing vs. retransmission scheduling.

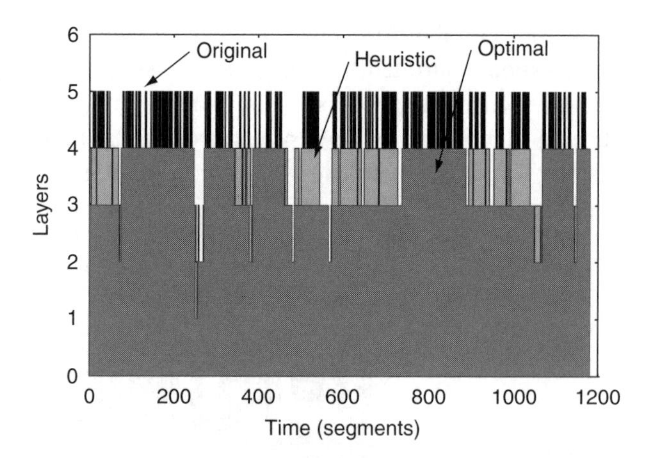

Figure 6.4 Comparison of originally cached and polished (heuristic and optimal) video object.

due to polishing. A reduced spectrum (28 compared to 193 of the originally cached object) indicates a better perceptual quality compared to that of the originally cached object. The effect of *over-polishing* has been avoided. The example also shows the storage space that could be freed on a cache, based on the information that is gained by polishing. In comparison, for the simple heuristic that simply drops the top layer, the amount of storage space that can be freed through optimal polishing is higher. A simulative investigation on polishing as a mechanism for cache replacement is given in section 6.5.3.

6.3 EXISTING WORK ON POLISHING

Rejaie *et al.* [115] introduce a fine-grained cache replacement mechanism that allows the deletion of single segments. Each layer of a video is regarded separately. Beginning at the top layer of a video, for each single layer segments are removed from end to beginning, while in the case of polishing the whole video is regarded for the removal of segments. Thus, in the case of Rejaie's approach, segments of the top layer of the cached video are removed until none of the segments of this layer is left. If more space on the cache is needed, this process will be continued on to the next lower layer. In the case of polishing, segments from all available layers can be removed independently. The approach presented by Paknikar *et al.* [116] allows only the removal of complete layers which is somewhat similar to the heuristic presented in section 6.5. Also in the analytical investigation performed by Kangasharju *et al.* [117] only complete layers can be dropped.

Quality-based caching [153] is an additional approach for partial caching which assumes that metadata information about the quality of a scalable video is available. For example, the metadata would provide that removing the top layer of a 5-layer video would reduce the quality of the video by 20%. The authors leave open how this necessary metadata can be obtained.

6.4 OPTIMAL POLISHING

Polishing a layer-encoded video means to minimize the spectrum while at the same time maximizing the number of segments played out to the client and could thus be regarded as a multi-objective optimization problem. Two characteristics of that optimization problem make it hard to be treated directly: on the one hand, the two competing optimization goals and, on the other hand, the nonlinear quadratic form of the spectrum. Therefore, the spectrum is substituted by a new metric, namely layer variations, and a utility-based approach where parameters for the relative weighting between the two competing goals of polishing are introduced.

Polishing – which under these prerequisites means maximizing the playback utility of a video – can be formulated as the mixed integer programming problem [154] given in Figure 6.6.

The two parameters u_l and p describe the utility of the video playout. u_l is the utility for receiving layer l (and all lower layers) in one period t. Obviously, the more layers are played back the higher the utility. p describes the utility loss for a layer change. By including u_l into the optimization process the *over-polishing* effect described in section 6.2.3 is avoided. p prohibits quality loss by changing the number of layers too often.

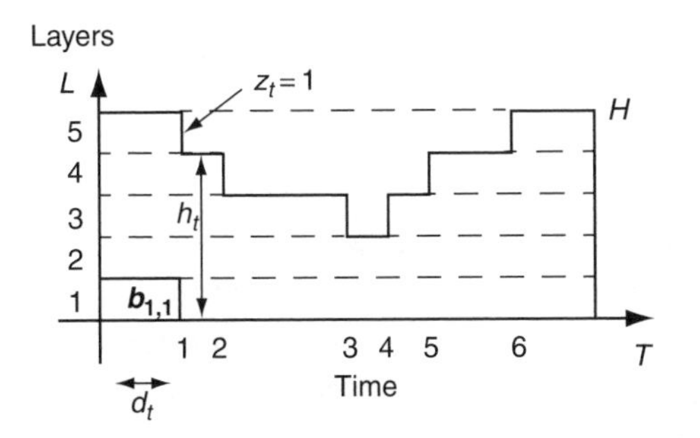

Figure 6.5 Cached layer-encoded video.

Indices:

$l = 1, \ldots, L$ – layer of the video
$t = 1, \ldots, L$ – period t

Parameters:

h_t^{cached} – number of the highest layer that is cached for period, t, all lower layers are cached in period, t, too
H – sufficiently large number ($H \geq L$)
d_t – length of period (in seconds)
u_l – utility of receiving layer, l, for one second (if video is played back for one second on layer 3 it generates a utility of $u_1 + u_2 + u_3$)
p – utility loss for a change in the number of layers that are played back

Variables:

h_t – the layer the video is played back in period, t
z_t – binary variable, one if a layer change occurs at the beginning of period, t, zero otherwise
b_{tl} – binary variable, one if video is played back in period, t, at layer, l, or higher, zero otherwise

Optimization problem:

$$\text{Max}\left(\sum_{l,t} u_l d_t b_{tl} - \sum_t p z_t\right) \quad (10)$$

subject to

$$h_t - h_{t-1} \leq H z_t \quad \forall t = 2, \ldots, T \quad (11)$$

$$h_{t-1} - h_t \leq H z_t \quad \forall t = 2, \ldots, T \quad (12)$$

$$0 \leq h_t \leq h_t^{cached} \quad \forall t = 1, \ldots, T \quad (13)$$

$$l b_{tl} \leq h_t \quad \forall t = 1, \ldots, T \quad \forall l = 1, \ldots, L \quad (14)$$

$$b_{tl} \in \{0, 1\} \quad \forall t = 1, \ldots, T \quad \forall l = 1, \ldots, L \quad (15)$$

$$z_t \in \{0, 1\} \quad \forall t = 2, \ldots, T \quad (16)$$

Figure 6.6 Optimization model.

The variable, h_t, contains the highest layer of the polished video at time, t; it can never be higher than the highest cached layer (see constraint (13) and Figure 6.5). The binary variable, z_t, is needed to account for layer changes in the target function. z_t is forced to one by constraints (11) and (12) when the highest layer of the polished video changes. The binary variable, b_{tl}, stores whether a layer, l, is included in the polished video in period t or not, constraint (14) expresses its relationship with the highest layer, h_t.

This problem can be solved with standard techniques much as branch and bound and the Simplex algorithm [154]. Here, the commercial mathematical programming solver Ilog CPLEX [155] was used to solve the problem.

6.5 SIMULATIONS

To verify whether polishing is a valid approach and to obtain further information on the influence of the utility factors, u_l and p, a series of simulations were performed. An additional goal was to investigate also how a simple heuristic performs in comparison to the optimal polishing algorithm. This heuristic simply drops one or more adjacent layers completely, beginning from the top.

The simulations are performed in the following manner. For each simulation an instance of a layer-encoded video on the proxy cache is randomly generated as described in section 5.5. A discrete simulation time is used where one unit of time corresponds to the transmission time of a single segment. In Figure 6.4 (Original), an example video instance generated in this way is given. On each instance of a layer-encoded video created as described above the polishing algorithm is performed. The polishing algorithm was implemented using the mathematical programming solver Ilog CPLEX [155]. Before and after polishing the spectrum of the video is calculated in order to obtain information about the quality change. An example of such a simulation step is shown in Figure 6.4 (before and after polishing) with the following set of parameters: $u_1 = 1$, $u_2 = 1$, $u_3 = 1$, $u_4 = 1$, $u_5 = 1$ and $p = 8$.

6.5.1 Utility Parameters

To obtain better insight into the influence of the parameters u_l and p, a series of simulations with varying values for those parameters was performed. The results of this simulation are presented in Figures 6.7 and 6.8. For each parameter set, 100 video objects were randomly created and polished as described above. The average spectrum and the average total number of segments were calculated before and after running the polishing algorithm. Figure 6.7 shows the results for three different simulations and the spectrum for two versions of the heuristic. In the first version the top (Heu(1)) layer is dropped and in the second the two top (Heu(2)) layers are dropped. Then the average spectrum and average number of segments of all 100 resulting objects are calculated. As can be derived from both figures, the heuristic has the disadvantage that it is static while, in the case of optimal polishing, the selection of the parameters u_l and p influences spectrum and number of segments of the polished video object. On the other hand, the heuristic is simple and can be applied with little computational effort and the results obtained are fairly close to those of the optimal polishing.

In the case of polishing, the different simulations were performed with $u_l = 1$, $u_l = 1/l$ and $u_l = 1/l^2$, respectively.

In this specific simulation the heuristic turns out to be an alternative compared to the optimal polishing. A closer look at Heu(2) shows that the spectrum is significantly reduced while the amount of segments is reduced by a third. This result can also be achieved by applying the optimal polishing algorithm but at the price of a higher computational effort.

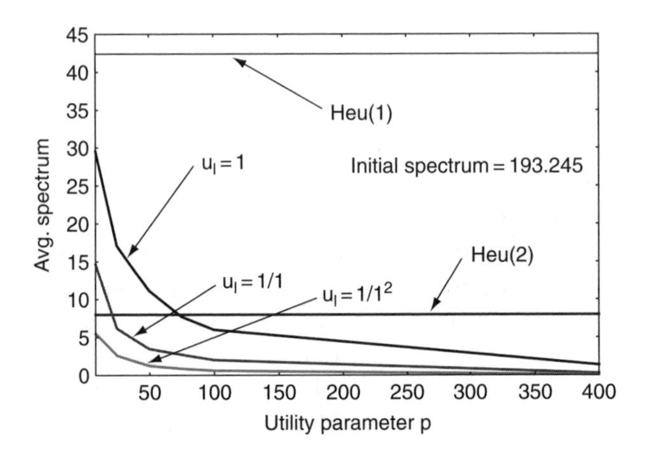

Figure 6.7 Average spectrum.

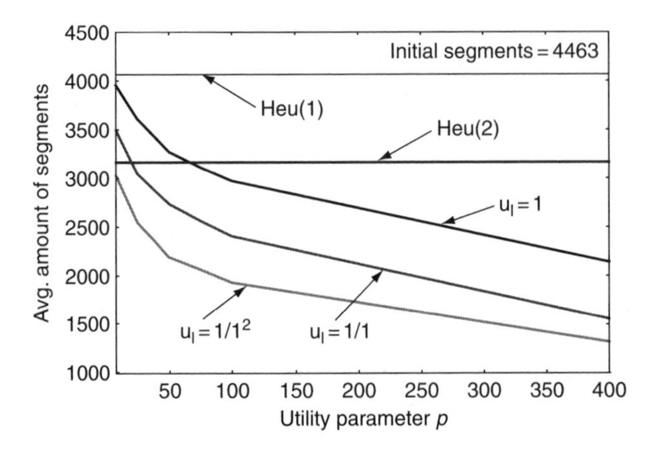

Figure 6.8 Average number of segments per video object.

An additional series of simulations was performed with the difference that the initial videos used as input for the polishing simulation were generated in a different way. In this case TFRC traces generated by a ns2 simulation were used to generate the initial video. The bandwidth information generated by the ns2 simulation is used to determine how many layers of a video can be transmitted during a certain period. This complies with a cached layer-encoded video that has been transmitted via TFRC. An example transmission is shown in Figure 7.1. This procedure is described in more detail in section 7.2.3.

Figure 6.9 depicts the resulting average spectrum obtained by the application of the heuristic and the optimal polishing algorithm. Compared to the results of the simulation where the videos are generated randomly, the results of optimal polishing are similar. Comparing the results for Heu(2) (Figures 6.7 and 6.9) with each other shows that the reduction in the spectrum is not as high with the TFRC-based simulation. This difference is caused by the fact that TFRC also starts with a slow-start (identical to TCP) and, for a short while until the occurrence of a loss event, the sending rate can become quite high. Thus, reducing the top or the two top layers might affect only a small portion of the video. In simulations presented in the following section it is shown that this static behaviour of the heuristic can reduce the efficiency of the heuristic.

The difference in the average number of segments for the random (Figure 6.8) and TFRC-based (Figure 6.10) simulation is caused by the fact that the TFRC traces are shorter in duration. The length of such a trace is equivalent to 400 time units while in the case of random-based simulation

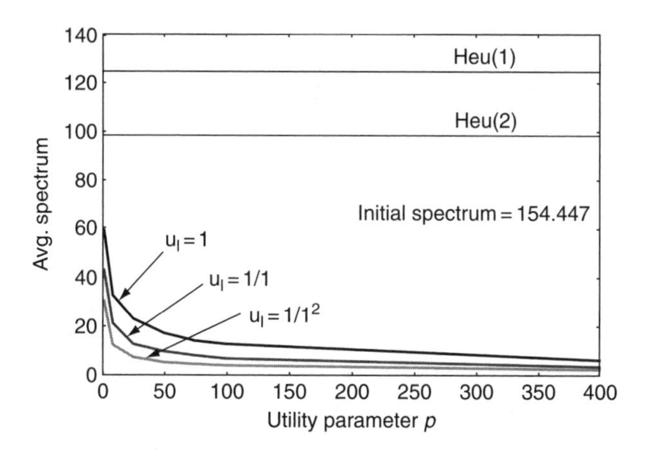

Figure 6.9 Average spectrum (TFRC-based).

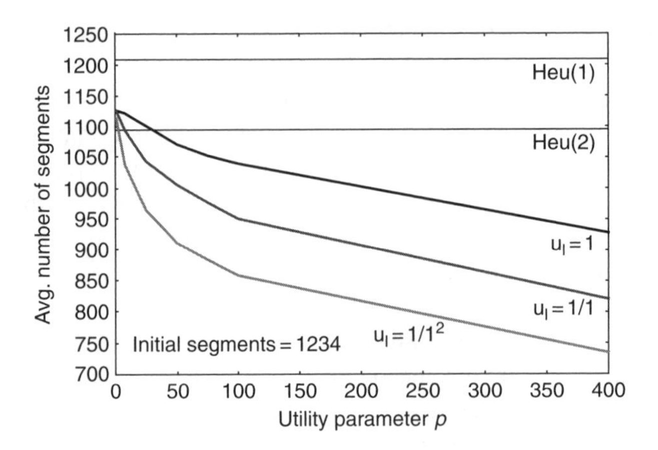

Figure 6.10 Average number of segments per video object (TFRC-based).

the layer-encoded video has a length of approximately 1200 time units. Nevertheless, the behaviour of the optimal polishing against u_l and p is almost similar for both simulation types. The average number of segments decreases with an increasing p and a decreasing u_l.

6.5.2 Polishing Layer-encoded Video with Different Quality Regions

An example of a video object that consists of two different quality regions, is shown in Figure 6.11. There can be several reasons for the creation of such a video object on the cache. One possibility is a limited bandwidth between server and cache caused by competing traffic. In the given example the transmission of the competing traffic started after half of the video was already streamed from the server to the cache. The occurrence of two major quality regions in a cached video is rather arbitrary, since several such *regions* with different quality levels can occur, depending on the situation on the path between server and cache. Nevertheless, the chosen example is sufficient to demonstrate the drawback of the heuristic presented above. As can be seen in Figure 6.11(b), the disadvantage of the heuristic is the fact that only the region with the better quality (higher number of layers) is polished. In this example the two top layers are dropped. This effect does not occur with optimal polishing where both regions are polished. Applying the heuristic might be annoying for the viewer. The already existing quality decrease between the two regions is even intensified by the high

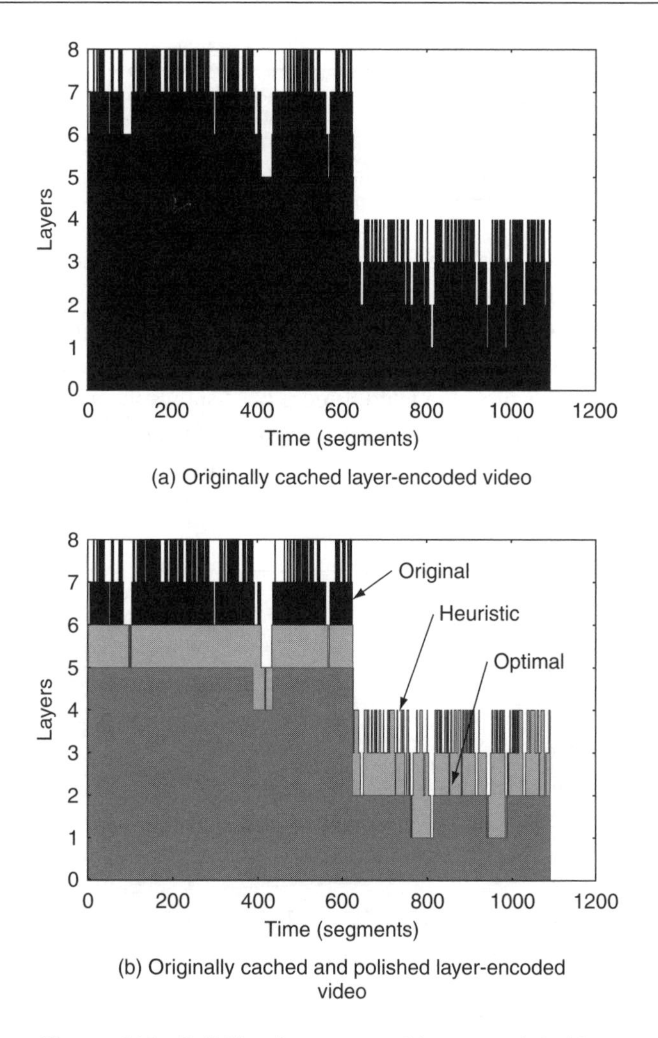

Figure 6.11 Polishing for two-staged layer-encoded video.

number of quality changes in the second half of the video. With optimal polishing, quality variations are reduced in both regions and, thus, the quality decrease between the two regions is not that intense. Comparing the two resulting spectra (see Table 6.1) of the second region demonstrates the effect mentioned above. The spectrum in the second region is significantly higher for the heuristic than for optimal polishing, while for the first region both spectra are identical.

For the case of the layer-encoded video consisting of two quality regions a series of simulations was also performed. The simulation environment is

Table 6.1 Spectrum per region

	Region 1	Region 2
Heuristic	6	86.81
Optimal polishing	6	1.2

identical to the one described in section 6.5 with a slightly modified creation process for the initially cached video that results in quality variations as shown in Figure 6.11(a). Similar to the simulations described in section 6.5.1, 100 video objects are randomly created and the average spectrum and total number of segments are calculated. Figure 6.12 shows the average spectrum for the optimal polishing depending on the parameters, u_l and p. The resulting spectra for the three variations of the heuristic are always larger than those for optimal polishing.

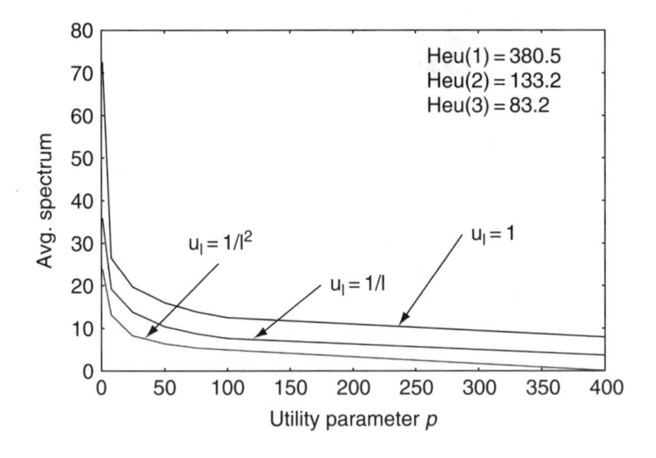

Figure 6.12 Average spectrum.

The comparison of the average number of segments per video object (see Figure 6.13) shows that the three variations of the heuristic result in higher values than the optimal polishing. An interesting case is the one for $u_l = 1$ and $p < 50$ where the spectrum for the optimal polishing is lower than the one of Heu3 but the number of segments is larger. This is a different result to the one presented in section 6.5.1 where the spectrum for the optimal polishing is always higher than the one for Heu2, if the number of segments is larger for optimal polishing than for Heu2. The result of this simulation shows the advantage of the optimal polishing compared to the heuristic in the case where the cached video consists of regions with varying quality

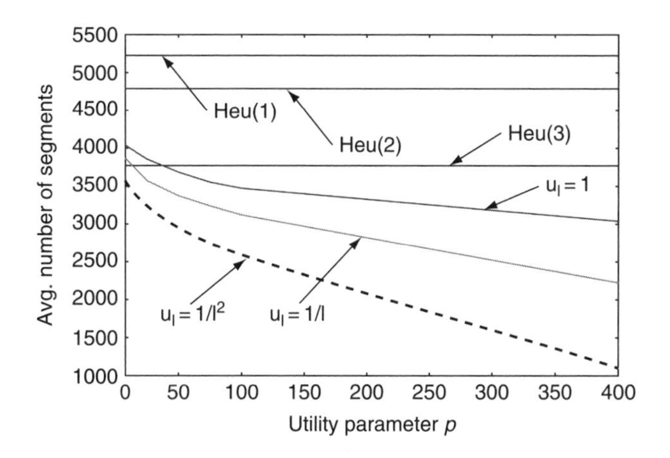

Figure 6.13 Average number of segments per video object.

levels. This advantage becomes even more obvious if one imagines cached video objects which consist of more than two regions with different quality levels. The polishing effect (i.e. reducing the amount of layer variation) would probably only occur in the region with the highest quality level, while the remaining regions would remain unpolished.

6.5.3 Cache Replacement

The goal of an additional set of simulations was to investigate how polishing can be used for cache replacement. This simulation is thought to demonstrate how useful polishing can be for cache replacement. In section 6.5.4 an additional simulation is presented that also takes popularity information about the single video objects into account for the cache replacement.

At the beginning of the simulation it was assumed that the cache was initially filled with 50 unpolished video objects, consuming all of the cache's storage space. Then, 50 additional objects should be incrementally, i.e., one per time slot, stored on the cache. To generate the additional storage space for these objects in the first step the already cached objects are polished. If the space gained by polishing is not sufficient or in the case that all objects are already polished, a cached object will be removed in FIFO manner. Figure 6.14 gives an overview of how this simulation was performed.

This simulation was performed for the parameter set $p = 50$ and $u_l = 1$. In an additional simulation (no-polishing), objects are not polished but simply removed. The cache replacement simulation was also performed for the two versions of the heuristic. Figure 6.15 depicts the spectrum for each of the

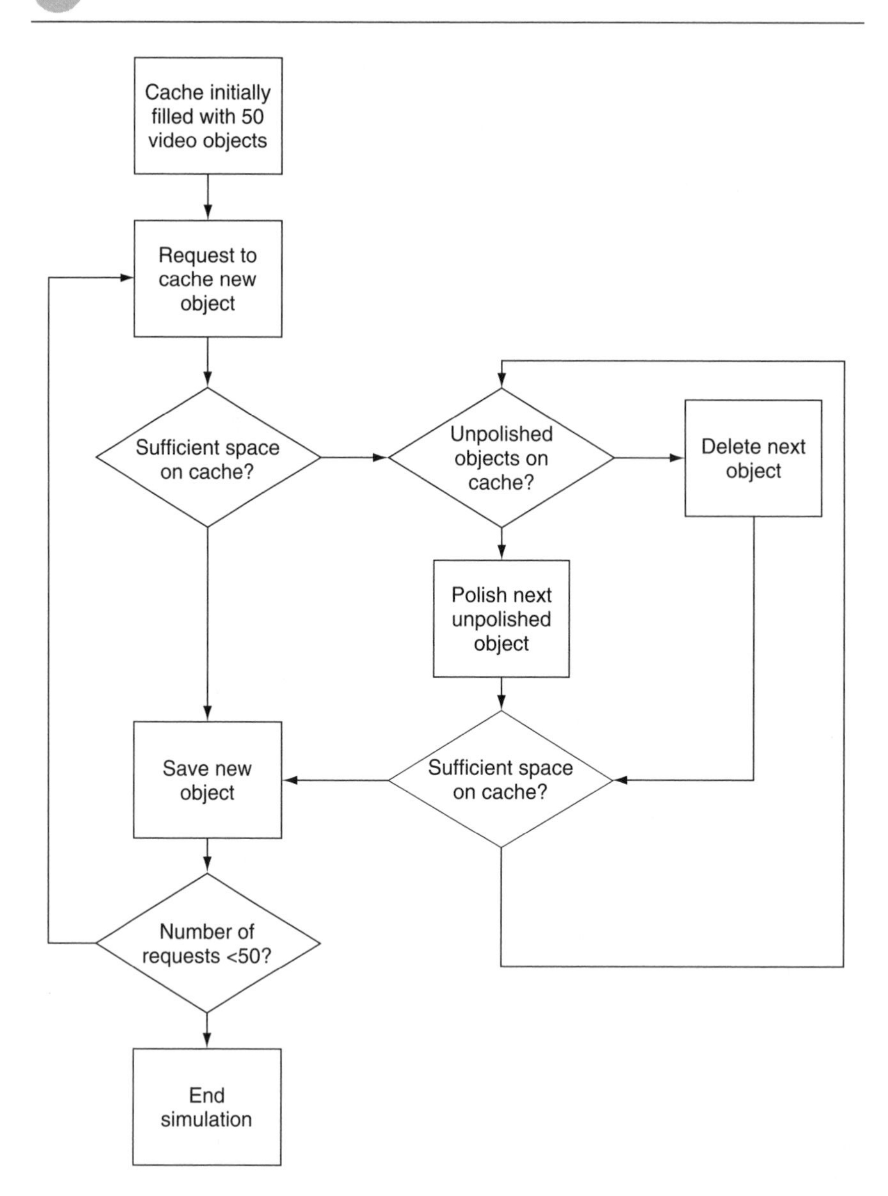

Figure 6.14 Simulation procedure.

simulations, while the average number of segments per object is shown in Figure 6.16. An interesting result is revealed by the comparison of the two graphs for the case $p = 50$ and $u_l = 1$, where a high reduction of the spectrum results in a moderate reduction of the average number of segments per cached object. Compared to a cache replacement which does not incorporate

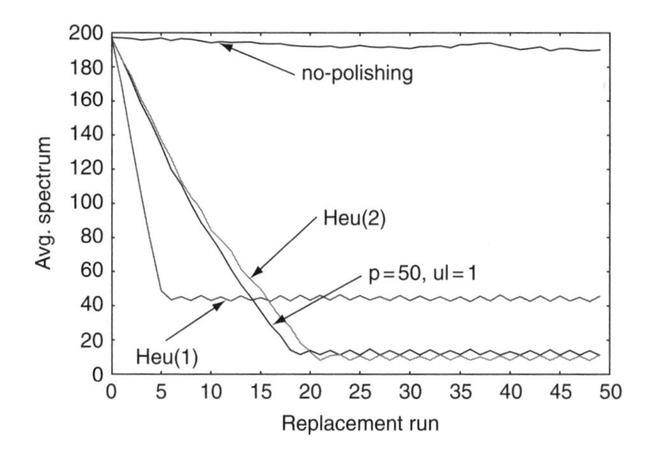

Figure 6.15 Spectrum for cache replacement.

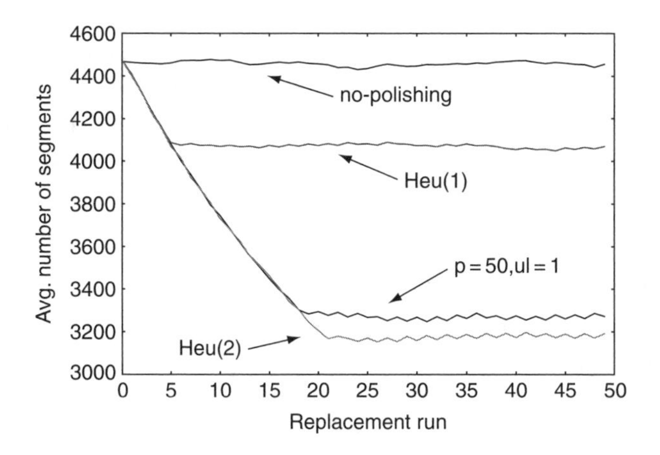

Figure 6.16 Average number of segments per object for cache replacement.

polishing the total amount of video objects stored on the cache is higher in the polishing simulations (see Table 6.2). Thus, integrating polishing into the cache replacement method is beneficial since a higher number of video objects can be cached and layer changes are reduced (smaller spectrum). Based on the parameters that can be chosen for the polishing method, the behaviour of the cache replacement method can be influenced. For example, a cache operator can control, with the aid of these parameters, whether more objects in a lower quality or vice versa should be cached. Also in this simulation, the results of the heuristic are close to the optimal polishing result.

Table 6.2 Total number of objects

Simulation	no-polish	$p=50$, $u_l=1$	Heur(1)	Heur(2)
Objects	50	68	55	70

6.5.4 Popularity-based Cache Replacement

The goal of this simulation is to investigate whether optimal polishing can be applied to polish cached videos based on their popularity. That means, fewer segments are deleted from popular videos while the number of deleted segments increases for less popular objects. Thus, the quality of the cached object is directly related to its popularity. In contrast to the aforementioned approaches, polishing is not performed individually, but the complete content of the cache is regarded and polished according to the popularity of each single object and the amount of space that should be freed. Similarly to the problem presented in section 6.4, this problem can be formulated as a mixed integer programming problem as shown in Figure 6.17.

Compared to the model presented in section 6.4, two additional parameters are introduced: the popularity of each video object, w_v, and the total capacity of the cache's storage that the already cached objects can consume, K^{max}. The latter allows one to determine how much storage space should be freed on the cache to allow the caching of new video objects.

The simulation for this investigation was changed in comparison to the one presented in section 6.5.3. Here, one can specify the amount of cache space that should become available for the caching of new data, $K^{total} - K^{max}$. In addition, each video object is assigned a certain popularity w_v. Figure 6.18 shows the originally cached and the resulting polished video object for the video with the highest (a) and the lowest (b) popularity on the cache. In this case, 10 video objects are stored on the cache and 25% of the total cache space is freed by polishing the cached videos according to their popularity. For the case of the most popular video object 20% of the original segments are deleted, while for the least popular video object 31% of the original segments are removed from the cache.

This simulation was performed 20 times. For each single simulation the initial cache state was randomly generated. Table 6.3 shows the percentage of segments that were removed from each of the 10 cached video objects. Objects which are shaded equally were assigned the same popularity value. The popularity is highest for objects 1, 2 and 3 and is lowest for objects 7, 8, 9 and 10. The popularity for 4, 5 and 6 lies in between the other two groups. The results of this simulation show that with the extended polishing algorithm a very fine granular cache replacement can be achieved. With this

Indices:
$v = 1, \ldots, V$ – number of video object
$t = 1, \ldots, T$ – period t

Parameters:

h_t^{cached} – number of the highest layer that is cached for period, t, all lower layers are cached in period, t, too.
H – sufficiently large number ($H \geq L$)
K^{max} maximum capacity available for already cached vedio objects
w_v – popularity of the video object
u_l – utility of receiving layer, l, for one second (if video is played back for one second on layer 3 it generates a utility of $u_1 + u_2 + u_3$)
p – utility loss for a change in the number of layers that are played back

Variables:

h_t – the layer the video is played back in period, t
z_t – binary variable, one if a layer change occurs at the beginning of period, t, zero otherwise
b_{tl} – binary variable, one if video is played back in period, t, at layer, l, or higher, zero otherwise

Optimization problem:

$$\text{Max} \sum_{v} \left(\sum_{l} \sum_{t} w_l u_{lt} b_{vtl} - \sum_{t} p z_{vt} \right) \quad (17)$$

subject to

$$h_{vt} - h_{vt-1} \leq H z_{vt} \quad \forall t = 2, \ldots, T \; \forall v = 1, \ldots, V \quad (18)$$

$$h_{vt-1} - h_{vt} \leq H z_{vt} \quad \forall t = 2, \ldots, T \; \forall v = 1, \ldots, V \quad (19)$$

$$0 \leq h_{vt} \leq h_{vt}^{cached} \quad \forall t = 1, \ldots, T \; \forall v = 1, \ldots, V \quad (20)$$

$$l b_{vtl} \leq h_{vt} \quad \forall t = 1, \ldots, T \; \forall v = 1, \ldots, V \; \forall l = 1, \ldots, L \quad (21)$$

$$b_{vtl} \in \{0, 1\} \quad \forall t = 2, \ldots, T \; \forall v = 1, \ldots, V \; \forall l = 1, \ldots, L \quad (22)$$

$$z_{vt} \in \{0, 1\} \quad \forall t = 2, \ldots, T \; \forall v = 1, \ldots, V \quad (23)$$

$$\sum_{v} \sum_{t} b_t h_{vt} \leq K^{max} \quad \forall t = 1, \ldots, T \; \forall v = 1, \ldots, V \quad (24)$$

Figure 6.17 Cache replacement optimization model.

algorithm it is possible to free cache space for new content while data from already cached content is removed according to the popularity of the content. In this specific example 25% of the caches' storage space is available for the caching of new content while none of the cached objects had to be removed completely.

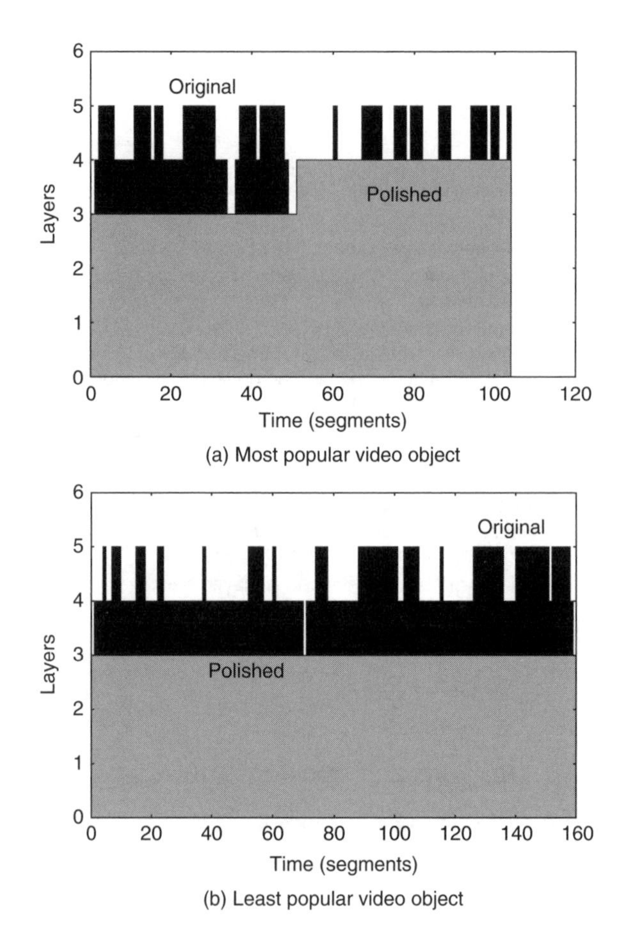

(a) Most popular video object

(b) Least popular video object

Figure 6.18 Polishing for most and least popular video object.

Table 6.3 Average number of removed segments

Video object	1	2	3	4	5	6	7	8	9	10
Average number of removed segments (%)	18	14	18	26	26	25	32	31	32	31
Avg. spectrum of polished object	5.6	4.3	5.3	3.1	4.0	5.4	2.2	2.0	2.8	2.4
Avg. spectrum of unpolished object	18.1	19.3	19.9	23.3	32.2	26.7	23.8	21.3	23.9	25.8

6.6 SUMMARY

In this chapter, a new technique, called polishing, is presented which can be applied either for the streaming of data from the cache to the client or as a cache replacement mechanism. Polishing makes use of the fact that a reduction in layer variation can also be achieved by not transmitting certain segments, although this means that some of the data available at the cache is omitted from the client. Optimal polishing, which means to maximize the playback utility of a video, is formulated as a mixed integer programming problem. Since optimal polishing is dependent on several parameters, a simulative investigation is performed to gain better insight into the influence of these parameters. In this simulation, the results of optimal polishing are compared with a simple heuristic that drops certain layers completely. The results show that the heuristic can, in specific cases, achieve similar results to optimal polishing with less computational effort. But simulations with layer-encoded video that consists of different quality regions demonstrate the drawback of the heuristic. In addition to the parameter analysis, two supplementary simulations on cache replacement were performed. The first one is rather simple and does not take the popularity of single video objects into account. With this simple cache-replacement method one can store a higher overall number of objects on the cache while, in parallel, layer variations are reduced. For the second kind of simulation of cache replacement, the optimal polishing mixed integer problem is extended to vary the intensity of polishing based on the popularity of the video object. The results of this simulation show that, by applying the extended optimal polishing, storage space on the cache can be freed while the number of segments is reduced according to the popularity of the single objects. In short, polishing is a valid means of reducing variations in layer-encoded video, either if retransmission from the cache cannot be performed, or as a cache replacement mechanism.

7

Fair Share Claiming

7.1 MOTIVATION

In Chapter 5, different retransmission scheduling algorithms that meet users' demands to watch high-quality video with relatively little quality variation were developed and compared with each other. In this chapter, the focus is on how these scheduled retransmissions can be combined with a TCP-friendly transmission method by claiming a fair share for the TCP-friendly session. Transmitting a layer-encoded video in a TCP-friendly manner does not always result in the case that the session claims its fair share of network resources, as will be shown in section 7.2. Therefore, a mechanism is proposed, called fair share claiming (FSC), which combines the transmission of a layer-encoded video and some additional data, resulting in the utilization of the fair share a session is entitled to. The applicability of FSC is investigated on the basis of a simulation and is compared with the results from Chapter 5, while the scheduling heuristics from Chapter 5 were also applied. Additionally, an implementation design for FSC, that makes use of already existing protocols for video streaming, is given.

7.2 PERFORMING TCP-FRIENDLY STREAMING IN COMBINATION WITH RETRANSMISSIONS

In this section, a fair share claiming mechanism that makes use of the additional bandwidth that is not claimed by the layer-encoded video, without breaking the cooperative rules implied by TCP's resource allocation model, is presented. After a detailed description of the FSC mechanism in section 7.2.1

Scalable Video on Demand: Adaptive Internet-based Distribution M. Zink

existing work on FSC is presented in section 7.2.2. The simulation environment for FSC is presented in section 7.2.3 and finally the simulation results are shown in section 7.2.4.

7.2.1 TCP-friendly Streaming

First of all it should be mentioned that is was not the goal of this work to develop new TCP-friendly mechanisms for streaming. In recent years several protocols for the transport of non-TCP traffic with TCP-friendly congestion control have been developed. An overview of these protocols is given in section 3.4. For the work on FSC it was decided to use TCP-friendly rate control (TFRC) [47] for several reasons.

TFRC is a rate-based congestion-control protocol with good TCP-friendliness. Its main advantage in combination with A/V streaming is the fairly smooth rate in the steady-state case and, therefore, applications that rely on a fairly constant sending rate are supported. In addition, the protocol is end-to-end, which does not require any modifications to the network infrastructure. Transmitting a layer-encoded video with the maximum rate that TFRC would allow does not always make sense. If, for example, the possible transmission rate were much higher than the actual rate needed for the video, the receiver might need a large buffer to store segments until their playout time was reached. Especially in the case of mobile receivers such as handhelds, buffer size might be a scarce resource. Since one of the major goals of the SAS architecture is to support the heterogeneity of the Internet, FSC is a mechanism well suited to fit that requirement, as is shown in the following. In Chapter 8, it is shown how FSC can be applied in the case of TCP-friendly transmission between server and cache and an uncontrolled transmission (pure UDP-based) between cache and client.

Rate changes in TFRC will not always result in a rate change for the layer-encoded video because the encoding format provides only a certain number of different layers, resulting in a finite amount of possible transmission rates. This can result in a situation where the actual possible transmission rate (determined by the TFRC algorithm) and the rate constituted by the sum of several layers might differ. The following example was chosen to illustrate the problem in more detail: for example, a layer-encoded video that consists of up to three layers, each requiring a constant transmission rate of 0.5 Mbit/s that should be transmitted in a TCP-friendly manner via TFRC. At a certain point in time during the transmission, the TFRC algorithm determines a maximum possible transmission rate of 1.3 Mbit/s. This would allow the transmission of two layers of the layer-encoded video, but it would not be possible to transmit an additional third layer. Provided that the video was not transmitted faster than was necessary, these would be 0.3 Mbit/s to spare. This additional bandwidth is the fair share that may be claimed by a corresponding TCP session, yet owing to the inelastic and discrete nature of layer-encoded video it cannot be claimed. Nevertheless, finding some data to fill this gap would allow the stream to claim its fair share without breaking the cooperative rules implied by TCP's resource allocation model. Figure 7.1 depicts an example TCP-friendly layered video transmission.

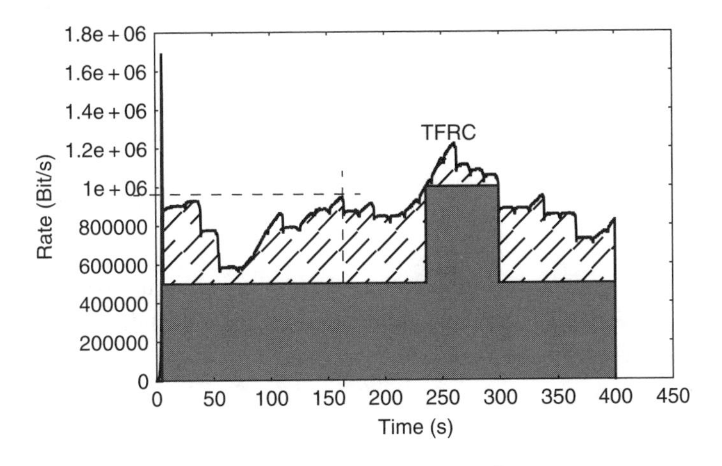

Figure 7.1 Example layer-encoded video transmission via TFRC.

The creation of the TFRC trace is explained in more detail in section 7.2.3. In this example scenario the transmission rate for the layer-encoded video is only increased in the case that the rate determined by the TFRC mechanism would allow the transmission of an additional layer. If this is not the case, the additional bandwidth (hatched in Figure 7.1) could be used for the transmission of additional data. In the example shown in Figure 7.1 that would be an additional 142 Mbyte. Here, the focus is particularly on the retransmission of missing segments of the video that is currently streamed or videos that are already (but not completely) stored on the cache to claim the fair share for this TCP-friendly session (see Figure 7.2). These techniques are referred to as *in-band* FSC for the former case and *out-of-band* FSC for the latter. In Chapter 5, different retransmission scheduling algorithms that could be used to determine which of the missing segments should be transmitted have already been devised and analysed. The following simulation should shed

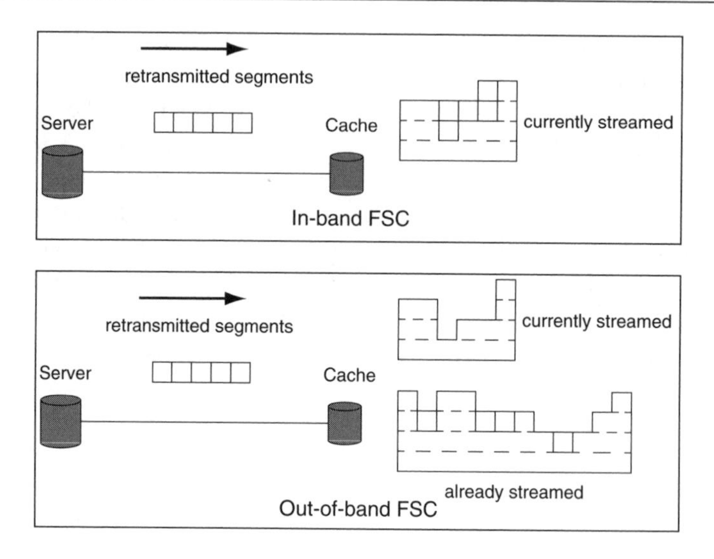

Figure 7.2 In-band and out-of-band FSC.

some light on whether the combination of the two techniques (retransmission scheduling and FSC) is an appropriate method for improving the quality of a cached layer-encoded video.

7.2.2 Existing FSC Approaches

Rejaie *et al.* [96] and Saprilla and Ross [99] also present mechanisms that claim their fair share and support the transport of layer-encoded video. Both assume that the client has sufficient buffer to allow a transmission rate higher than the receiver's consumption rate. While the first approach is limited to constant bit rate (CBR) encoding the second also supports variable bit rate (VBR) layered-encoding. In contrast to the FSC mechanism presented here, video transmission into caches is not considered. Neither mechanism supports the transmission of data that has already missed its deadline for timely consumption at the client and, therefore, neither offers any functionality to improve the quality of a video that is being cached or already stored on a cache.

Another approach that supports scalable streams is presented by Law *et al.* [156]. In their work the focus is mainly on server efficiency and scalability. In comparison to the FSC approach the quality is adapted owing to the capabilities of the receiving client rather than to network conditions.

Their architecture neither envisions caches nor incorporates TCP-friendly streaming.

Kangasharju *et al.* [157] consider the combination of caching and layered video, the latter only for the support of heterogeneous clients and not for congestion-control purposes. Furthermore, the emphasis of their work is on optimal cache replacement decisions viewed over *all* videos stored in a cache. For FSC, however, a two-stage decision process is assumed where in the first stage a video is selected for storage in a cache and then the retransmissions of missing segments are scheduled independently of the cache status of other videos. While this represents a restricted problem, it ensures that the overall problem still remains manageable. Another difference in their work is the fact that missing segments of a certain layer are only streamed directly to the client, in contrast to the FSC approach where the segments are transmitted to the cache to achieve a quality improvement for more than one client.

7.2.3 Simulation

The simulation is split into three single steps in order to create a scenario that represents the mechanisms presented in section 7.2.1 and to keep the simulation environment more generic:

(1) Creation of a TFRC trace.
(2) A possible layer-encoded video transmission that is derived from the TFRC trace.
(3) Determination of segments that can be retransmitted owing to spare bandwidth.

Each single step is explained in more detail in the following sections.

7.2.3.1 Creation of TFRC Traces

The TFRC traces are created with the aid of the network simulator ns-2 [158] because a TFRC model is already included and ns-2 allows us to create traces that can be used as a basis for the second simulation step. With the simulation configuration shown in Figure 7.3 a scenario for the distribution techniques described in section 7.2.1 is modelled. The simulation consists of two routers, two senders and two receivers. The routers, R1 and R2, are connected via a duplex link (L1) having a bandwidth capacity of 15 Mbit/s and a delay of 100 ms. This represents a scenario that consists of a server S1

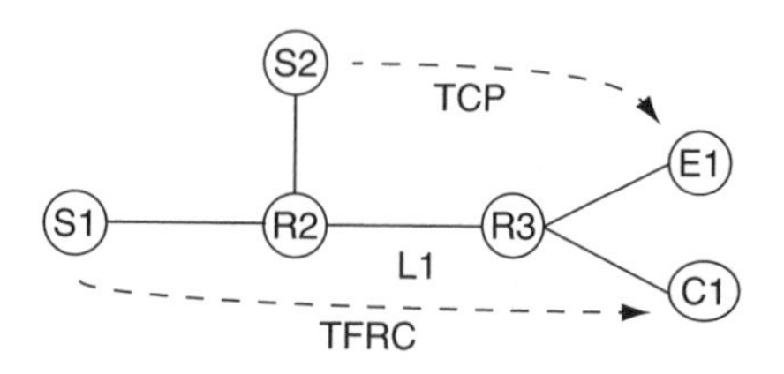

Figure 7.3 Simulation.

and a cache C1 that caches video streams and forwards them to requesting clients. To model competing web-like traffic between S2 and E1 we use ON/OFF sources as proposed by Floyd *et al.* [47]. In addition, there is a TCP session between S2 and E1 which serves as a reference in order to observe the TCP-friendliness of TFRC. It is active throughout the entire simulation. The ON/OFF sources are also enabled during the whole simulation. One long-lasting TFRC stream, representing the layer-encoded video transmission, is initiated at simulation start. A single simulation lasts for 400 seconds. The trace shown in Figure 7.1 was generated with the method described here.

7.2.3.2 Layer-encoded Video Transmission

In the simulations, layer-encoded video that can consist of up to three layers is assumed. In addition, all layers are of equal size and CBR encoded and, therefore, require an identical transmission rate, which is 0.5 Mbit/s for these simulations. To create a layer-encoded video transmission, a TFRC trace created by the methods described above is used as the starting point. A small C++ program was implemented that scans the bandwidth for each entry of the TRFC trace and determines the number of layers that can be transmitted based on the TFRC bandwidth. The rate for the layer-encoded video transmission in Figure 7.1 was generated in this way. It must be stated here that the strategy for increasing or decreasing one of the layers is very simple. In particular, a more intelligent strategy could also contribute to a smoother transmission of the video. During the execution of this program, an additional list is built to store information about the spare bandwidth that is available for the transmission of additional data. For the example shown in Figure 7.1 at 163.2 seconds a spare bandwidth of 452 000 bit/s would be determined. The simulation is discrete since the TFRC trace has a resolution of 0.2 seconds. This restriction had to be made to keep the overall simulation effort within reasonable limits. The error that is introduced by this simplification is negligible since the delay for the link between R1 and R2 is 100 ms and therefore the RTT is at least 200 ms, thus leading to the fact that two consecutive rate changes are never less than 200 ms apart.

7.2.3.3 Retransmission

To investigate retransmission scheduling in layered video caches in more detail a simulation environment was built and presented in section 5.5. In contrast to the simulations presented here, an instance of layer-encoded video was created randomly. The available bandwidth for retransmissions was constant for a single simulation and was only modified to compare the behaviour of the retransmission scheduling algorithms in relation to different amounts of available bandwidth. For the simulations presented in this section the bandwidth for retransmissions can change at each step of the simulation. For this reason the simulation environment had to be changed in two respects:

- For each step in the simulation the available bandwidth for the retransmission of missing segments must be calculated. This is performed with the aid of the list generated by the simulation tool described above that contains information about the available bandwidth for retransmissions at a certain point in time.
- If retransmissions are performed for the simultaneously streamed video (in-band) the retransmission scheduling algorithm can only consider the already transmitted part of the video. This is in contrast to our earlier work where we assumed that the complete instance of a cached layer-encoded video is known.

7.2.4 Simulative Experiments

Two different kinds of simulations as described in section 7.2.3 were generated, one for in-band and one for out-of-band FSC. The latter were performed to compare the unrestricted retransmission scheduling algorithm (see section 5.3) with the window-based algorithm presented in reference [115]. (For easier identification of both algorithms we refer to them as *unrestricted* and *restricted*, respectively.) Since the latter always looks a certain amount of time ahead of the current playout to determine segments for retransmission, it is not applicable for in-band FSC.

7.2.4.1 In-band FSC

The result of a single in-band FSC simulation is depicted in Figure 7.4. It shows how the quality of a layer-encoded video on a cache can be improved with the aid of the FSC technique. In this specific scenario it was possible to add an additional layer for more than half of the length of the complete video. Thus the next client requesting this video from the cache will have the chance to receive it in a significantly better quality than the first client. Unfortunately, there is a small gap for layer 2 between the 200th and 250th second that decreases the quality of the cached content. One possible way to close this gap would be to use the out-of-band FSC technique (during the transmission of some other video). To reduce the number of layer changes the caching strategy on the cache would decide to delete the short amount of the third layer that was cached owing to the peak of the TFRC around 5 seconds after the transmission started, for example, by applying polishing as presented in Chapter 6.

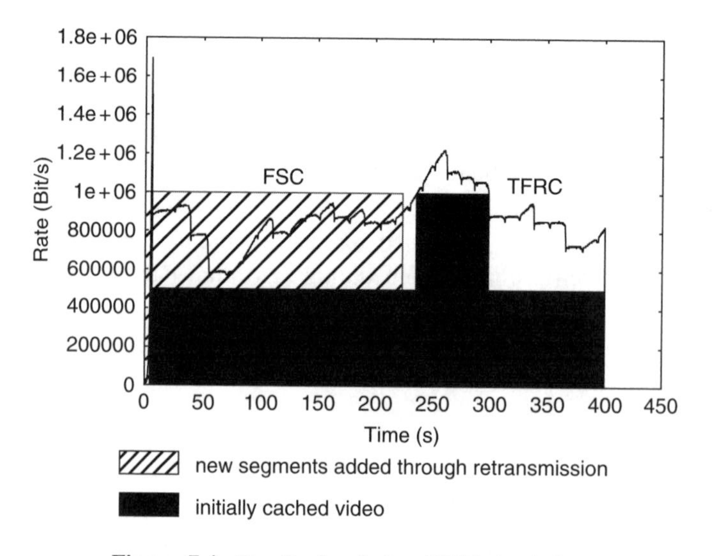

Figure 7.4 Result of an in-band FSC simulation.

For the complete in-band FSC simulation 100 different TFRC traces are created as described in section 7.2.3. For each one of these traces the additional available capacity for retransmissions is calculated. Thus, an average capacity of 5.15×10^8 bits is available, which is equivalent to one single layer 370 seconds long.

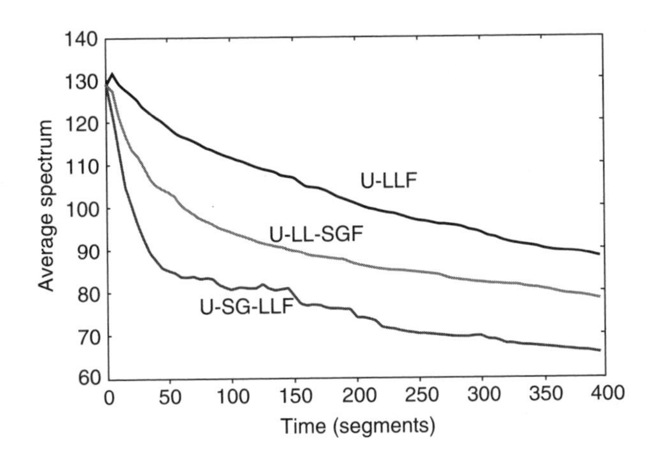

Figure 7.5 Average spectrum for 100 simulations.

Figure 7.5 shows the average spectrum for the three cache-centric heuristics. The outcome of this simulation confirms the results of the simulation presented in section 5.6 where a constant available bandwidth for the whole simulation was assumed. Again, U-SG-LLF is the best performing, followed by U-LL-SGF and U-LLF. An interesting detail is that all three spectra are monotonic decreasing. This is caused by the fact that in some cases only a small amount of bandwidth is available for retransmissions and, therefore, gaps will not be closed completely or, even worse, segments of layers that were not cached at all are retransmitted. The latter increases the number of layer changes, which leads to an increased spectrum. This short-term increase should be accepted to allow a quality improvement of the cached video in the long run.

7.2.4.2 Out-of-band FSC

The out-of-band FSC simulation was performed in a slightly different way from the in-band simulation. This is due to the fact that segments for an already cached video are retransmitted. To be able to compare the results of this simulation with the results of the in-band simulation, it is assumed that the cached video has the same layout as the layered video trace in Figure 7.4. That is, the initial transmission is identical to that of the in-band simulation, but without any retransmissions for this specific video. This allows comparison of the quality improvement between the in-band and the out-of-band technique in the unrestricted case. Since the 'layout' of the video is now completely known, both algorithms, restricted and unrestricted, can be applied. A second TFRC trace is generated which determines how much additional bandwidth is available for the retransmission of missing segments. The result of this simulation, which is depicted in Figure 7.6, clearly shows the disadvantages of the restricted algorithm. On account of the fact that only missing segments ahead of time from the actual playout point are regarded for retransmission, only small chunks of the missing segments can be retransmitted (the black boxes in Figure 7.6 only appear as boxes because of the low resolution of the plot). The problem of the restricted algorithm is shown in more detail in the magnified part of Figure 7.6 that represents an enlarged part of the out-of-band FSC simulation for this algorithm. The high frequency of layer changes is very annoying for the client currently watching the video (see the results presented in Chapter 4). In contrast to the restricted algorithm the result of the unrestricted algorithm is that the quality of the video is enhanced by one layer in one contiguous segment which, in this specific case, does not lead to additional layer changes compared to the initially cached video. The difference in the number of retransmitted segments in comparison to the in-band simulation (see Figure 7.4) results from the fact that the amount of spare bandwidth that is available for retransmission is higher in the in-band case. That is, the rates of the second TFRC trace allow the retransmission of a larger number of segments, as was the case for in-band FSC.

To exclude isolated effects that could occur by only performing a single simulation, 100 simulations were performed on the basis of the TFRC traces already created for the simulation described above. In Figure 7.7, the results

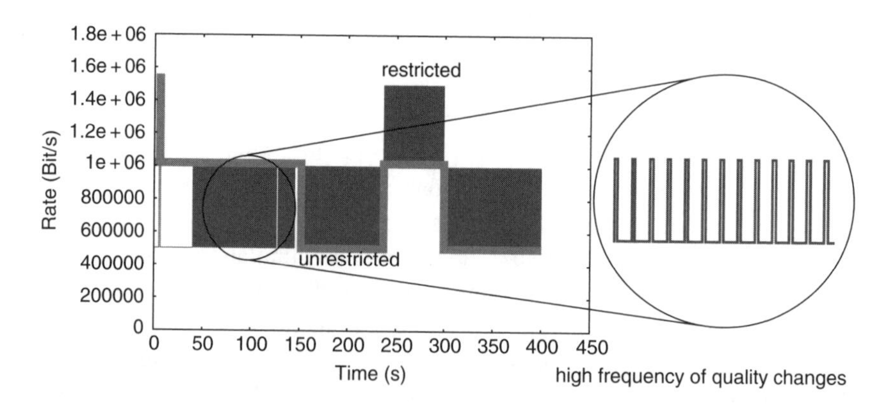

Figure 7.6 Out-of-band FSC simulation.

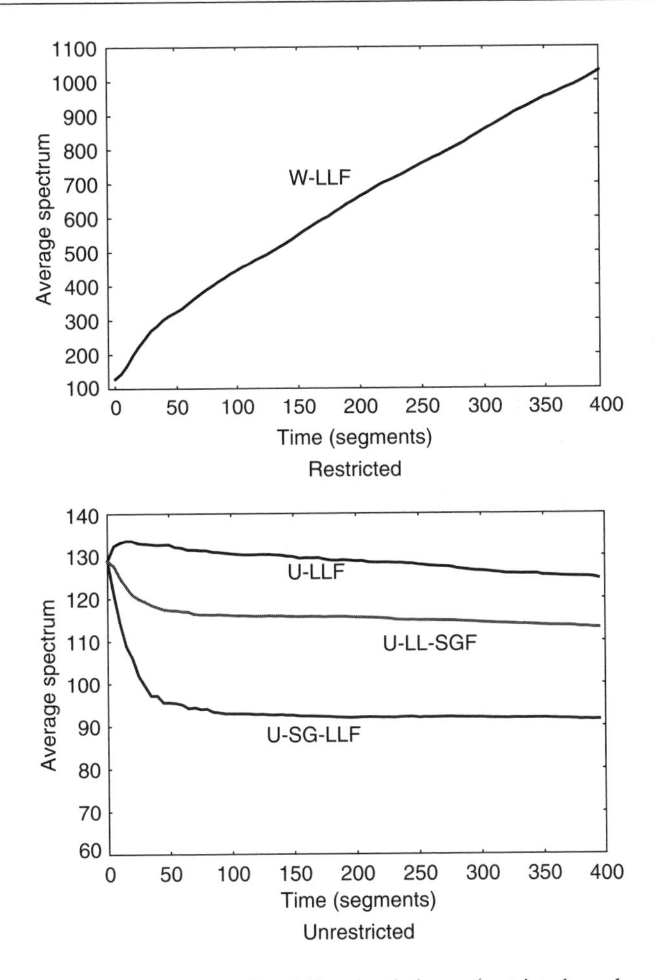

Figure 7.7 Average spectrum for 100 simulations (restricted and unrestricted viewer-centric).

of the average viewer-centric unrestricted and the restricted heuristics (see section 5.4) are shown. Here too the drawback of the restricted heuristic (W-LLF) becomes quite clear: the spectrum is constantly increased as a result of the effects shown in Figure 7.6. This is the only case where the results from the simulations presented in section 5.5.1 cannot be confirmed. Thus, the amount of additional bandwidth that is available for retransmissions influences the performance of the W-LLF heuristic. The results for the viewer-centric unrestricted heuristics confirm the results of the simulation presented in section 5.5.1.

7.2.4.3 Influence of the Total Number of Layers

It is obvious that the amount of bandwidth that is available for retransmission decreases with an increasing number of layers as the result of a higher adaptiveness to the bandwidth determined by TFRC. An additional simulation should investigate the dependency between the number of layers of a layer-encoded video and the resulting amount of bandwidth for retransmissions. In the simulation the same TFRC trace is used for each single step and the number of layers is varied between 2 and 20. Increasing the number of layers does not increase the maximum bandwidth of the layer-encoded video; rather the bandwidth of each single layer is decreased. This had to be done to make comparison of the results of each single simulation possible. Figure 7.8 presents the result of the simulations which shows the percentage of the overall capacity of the TFRC trace that can be used for retransmissions. The result states our assumption that the additional capacity will decrease with increasing layer granularity. Yet, even with a high number of layers there is still capacity for retransmissions available. The result of this simulation clearly shows that FSC is not applicable for FGS layer-encoded video because FGS allows one to adapt the enhancement exactly to the available bandwidth. An exception is the case of a thresholded FGS transmission, as described in section 4.7.2.

The correlation between the number of layers and the additional capacity for retransmissions is shown in Figure 7.9 for a single TFRC transmission.

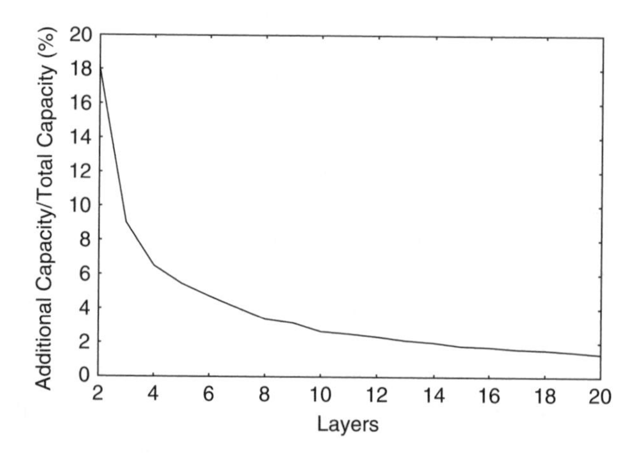

Figure 7.8 Relative additional capacity for retransmissions.

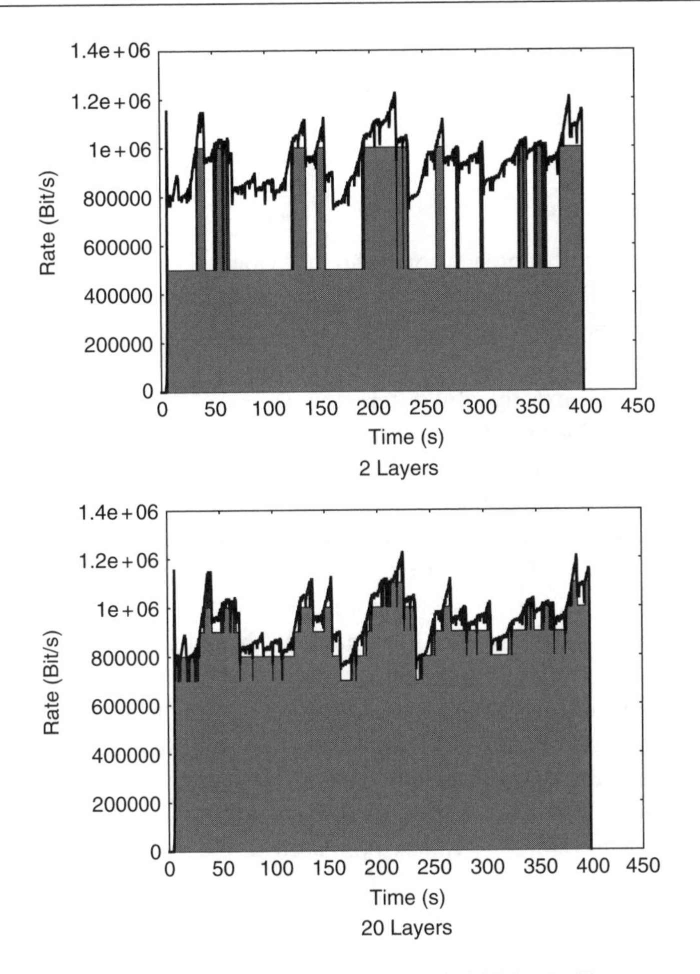

Figure 7.9 Comparison between 2- and 20-layer video.

7.3 IMPLEMENTATION DESIGN FOR FSC

In this section, the design for an implementation of the FSC technique which should be based on already existing, standardized and, if possible, well-established protocols and techniques is presented. However, this has not always been possible, mainly owing to the fact that the proposed TCP-friendly mechanisms require significant changes of the protocol design. Nevertheless, the design should only require modifications at server and cache to allow standard clients in this architecture. In general, a distinction between the transmission of time-critical data (the actual stream that is transmitted) and time-uncritical data (segments for retransmission) is made. It is the overall goal of this section to show that FSC can be reasonably integrated in streaming applications.

7.3.1 Protocol Suite

The most common approach for audio and video streaming is to use RTP[†] rather than UDP as transport protocols. It is well known that this approach lacks an appropriate congestion-control mechanism and may cause problems such as congestion collapse if the number of audio and video streams further increases. That is exactly why different variations of TCP-friendly protocols have been developed – to avoid the occurrence of such problems in the best-effort Internet. As mentioned above, TFRC is one of these protocols and in section 7.2.1 it is already stated why it is favoured as a TCP-friendly protocol for streaming environments. Another advantage is that the TFRC mechanisms can be integrated into the RTP protocol and, thus, the introduction of a completely new protocol in the streaming protocol suite is not necessary. This integration has the additional benefit that no modifications to UDP must be made and, therefore, possible kernel modifications can be avoided. To enable TFRC functionality in RTP, some new header information is needed (see Figure 7.10) and part of the overall protocol behaviour must be changed. Two of the additional header fields needed are already contained in the RTP header: *sequence number* and *time stamp*. The additional fields shown in Figure 7.10 must be placed in the RTP extension header.

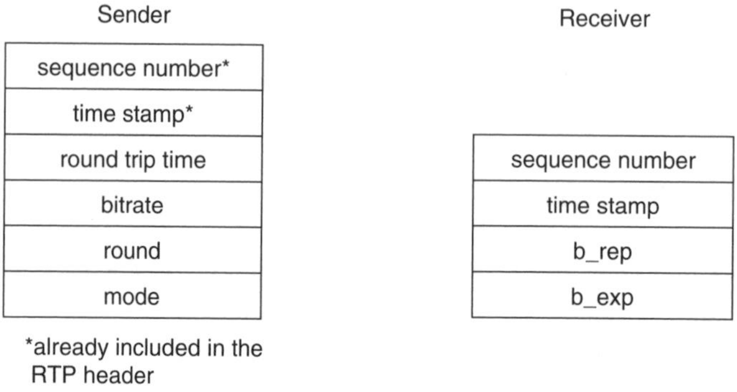

Sender

sequence number*
time stamp*
round trip time
bitrate
round
mode

*already included in the RTP header

Receiver

sequence number
time stamp
b_rep
b_exp

Figure 7.10 Additional TFRC header fields.

[†] For reasons of simplicity only RTP is mentioned; but always the combination of RTP and RTCP is meant.

The receiver reports needed by TFRC can be transported by the *application specific information* in the RTCP receiver reports. The frequency of RTCP receiver reports must be greatly increased since TFRC requires these reports to be sent every RTT. Since only unicast transmission is envisaged so far in the SAS architecture, the greater number of reports should neither restrict the raw data transmission nor cause an ACK implosion.

The identification of missing segments of a layer-encoded video that should be retransmitted is rather complicated. One might imagine that missing segments could be easily identified by the RTP sequence number, but this is not true in every case. The sequence number of an RTP packet would only be helpful if the data were stored as RTP packets on the server's disk, because the simple information of a sequence number would not be sufficient to identify the related part of, for example, a file that contains an MPEG-1 video where the packet length can vary (wire format and storage format do not necessarily have to be identical). In the case of LC-RTP (see Appendix A) the loss recognition is realized by a *byte count* which is included in each RTP header. The byte count represents the actual byte position of the data that is included in the RTP packet. Each server implementation has to transform the byte count value into its own file indexing information. As a consequence it is possible to have different file layouts on sender- and receiver-side. For example one server implementation stores the file as raw data and another stores some header information with it. A possible way of inserting the byte count into the RTP header and not into the payload is the use of the extension header of RTP. With the aid of the byte count, losses can be exactly identified: the receiver can maintain a list of losses; and the lost segments can be requested from the sender at another point in time. In section 8.4.1, an RTP payload format is presented that also allows the receiver to detect which segments of a layer have been lost or were not transmitted at all owing to network congestion. If lost segments should be retransmitted during the streaming session the RTCP application-specific header can be used to send the loss lists from receiver to sender. Should the retransmission be performed out of band, a TCP connection would be sufficient to transmit the loss lists to the sender. Since there now exist two cases that require an RTP extension header we propose that in the case of FSC the RTP protocol should be used with a modified extension header as shown in Figure 7.11.

An additional issue is the multiplexing of the initial video stream and retransmitted segments over one RTP session. In this case one can make use of RTP's mixing functionality. Originally this functionality was thought to combine RTP streams from different senders at a router into one RTP stream. Here, this functionality is used in a slightly different way: in this

```
 0                   1                   2                   3
 0 1 2 3 4 5 6 7 8 9 0 1 2 3 4 5 6 7 8 9 0 1 2 3 4 5 6 7 8 9 0 1
+-+-+-+-+-+-+-+-+-+-+-+-+-+-+-+-+-+-+-+-+-+-+-+-+-+-+-+-+-+-+-+-+
|      defined by profile       |            length             |
+-+-+-+-+-+-+-+-+-+-+-+-+-+-+-+-+-+-+-+-+-+-+-+-+-+-+-+-+-+-+-+-+
|                        round trip time                        |
+-+-+-+-+-+-+-+-+-+-+-+-+-+-+-+-+-+-+-+-+-+-+-+-+-+-+-+-+-+-+-+-+
|                            bitrate                            |
+-+-+-+-+-+-+-+-+-+-+-+-+-+-+-+-+-+-+-+-+-+-+-+-+-+-+-+-+-+-+-+-+
|                     loss measurement round                    |
+-+-+-+-+-+-+-+-+-+-+-+-+-+-+-+-+-+-+-+-+-+-+-+-+-+-+-+-+-+-+-+-+
|                             mode                              |
+-+-+-+-+-+-+-+-+-+-+-+-+-+-+-+-+-+-+-+-+-+-+-+-+-+-+-+-+-+-+-+-+
|                          byte count                           |
+-+-+-+-+-+-+-+-+-+-+-+-+-+-+-+-+-+-+-+-+-+-+-+-+-+-+-+-+-+-+-+-+
|                          byte count                           |
+-+-+-+-+-+-+-+-+-+-+-+-+-+-+-+-+-+-+-+-+-+-+-+-+-+-+-+-+-+-+-+-+
```

Figure 7.11 RTP header extension for congestion-controlled streaming.

scenario no physically separated senders exist but the layer-encoded video and the packets that should be retransmitted can be regarded as two logical sources. Thus both streams[†] can be transmitted via one RTP stream, whereas each stream is assigned a different *synchronization source identifier* (SSRC). This technique allows the RTP receiver to correctly identify each of the two streams and forward the packets to their correct destination. It might also be possible to mix more than two streams with this mechanism, but this is beyond the scope of this work. To identify each SSRC correctly the receiver needs additional information about the mapping between streams and SSRCs. The mapping information can be signalled to the receiver with the aid of the *private extension source description* (SDES) item of an RTCP source description packet. This type of RTCP packet contains a list of SSRCs and corresponding SDES items. The *private extension* item is meant for experimental or application-specific use. The SDES *private extension* consists of an *item identifier*, *length information*, *prefix length*, *prefix* and *value string* (see Figure 7.12).

```
 0                   1                   2                   3
 0 1 2 3 4 5 6 7 8 9 0 1 2 3 4 5 6 7 8 9 0 1 2 3 4 5 6 7 8 9 0 1
+-+-+-+-+-+-+-+-+-+-+-+-+-+-+-+-+-+-+-+-+-+-+-+-+-+-+-+-+-+-+-+-+
|     PRIV=8    |     length    | prefix length | prefix string ...
+-+-+-+-+-+-+-+-+-+-+-+-+-+-+-+-+-+-+-+-+-+-+-+-+-+-+-+-+-+-+-+-+
...             |                  value string                ...
+-+-+-+-+-+-+-+-+-+-+-+-+-+-+-+-+-+-+-+-+-+-+-+-+-+-+-+-+-+-+-+-+
```

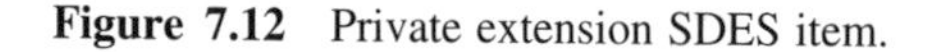

Figure 7.12 Private extension SDES item.

[†] To simplify description, the retransmission of segments is also described as a stream, although this is not technically correct.

The *prefix string* for this specific SDES item will be set to FSC to indicate that this information is related to the fair share claiming technique. For the *value string* three different strings are defined as in Table 7.1. This additional information allows the demultiplexing of the single sessions of an RTP stream and their correct assignments to instances for further data processing. In section 7.3.3, it is demonstrated how a correct data path could be established with the aid of the *Stream Handler* (SH) [159] architecture.

Table 7.1 Value string parameters

Value string	Description
STREAM	SSRC represents layer-encoded video stream
INBAND	SSRC represents a stream for retransmitted segments that belong to the in parallel streamed layer-encoded video
OUTBAND	SSRC represents a stream for retransmitted segments that belong to an already cached video

7.3.2 Retransmission Signalling

As mentioned above it might be possible to perform in-band and out-of-band retransmission with FSC. With out-of-band retransmission the respective video is already stored on the cache and one run of the retransmission scheduling algorithm should be sufficient to generate a retransmission list. A simple TCP transmission from receiver to sender to send the list of ordered missing segments should be sufficient. The sender stores this list and in the case of a retransmission request uses the list to obtain information about which segments of the original video should be retransmitted.

In the case of in-band retransmission the retransmission signalling must be handled in a different way. First of all the video is not entirely transmitted to the cache. The retransmission scheduling algorithm can only make decisions based on the already received part of the video. Thus the generated list of segments that should be retransmitted might change over time and updates of the list that exists at the sender must be performed. To be able to perform this update the initial list that is created by the retransmission scheduling algorithm should also be stored on the receiver. Each time the algorithm is performed again, the newly generated list and the stored list should be compared. If the differences reach a certain threshold value (for a suitable metric that measures similarity between loss lists) a new list must be transmitted to the sender.

7.3.3 Stream Handler Extension

Our experience with the implementation of streaming applications showed us the need for a generic architecture to handle continuous media streams. This became specifically clear during the development of our experimental KOMSSYS video streaming platform [159]. The platform is used for investigations on A/V distribution systems and, therefore, has to offer support for different encoding formats and transport protocols, but also distribution mechanisms under investigation. Such distribution mechanisms may combine unicast and multicast distribution or may apply segmentation and reordering for efficient delivery. During the initial implementation phase we quickly realized that a monolithic approach would not allow a simple integration of these new distribution mechanisms. This led to our decision to build an environment that is based on a stream handler (SH) architecture (see Appendix C).

In this section, the extensions are defined that must be made in the stream handler architecture to support the FSC technique. In the experimental VoD platform, server, cache and client make use of the stream handler architecture. Therefore, a great deal of stream handler modules have already been designed, implemented and tested. To support FSC in the streaming platform reusing these elements should be tried, and if necessary extending them or creating new stream handlers. First of all the modifications that have to be made to the already existing SHs are shown and then the design of the new SHs is presented.

7.3.3.1 RTP/RTCP and LC-RTP/LC-RTCP

The RTP and RTCP functionality is combined in the RTP SH. The FSC technique requires extensions and modifications to both RTP and RTCP as described in section 7.3.1. For the multiplexing and demultiplexing functionality the RTP SH must be able to receive data from more than one upstream SH (sender) and forward it to more than one downstream SH (sender) in the case of out-of-band retransmission. This is not the case for in-band retransmission, since the byte count information of LC-RTP defines the position of the retransmitted segment explicitly. Although only LC-RTP is used in the FSC case, the extension should be made for both protocols since RTP is a basis for LC-RTP and the new functionality may also be needed by RTP only. This allows a separated development of RTP with TFRC mechanism and LC-RTP.

On both the sending and receiving parts, extensions to provide TFRC functionality must be made, too. New fields must be added to the extension header as depicted in Figure 7.11. The changes for RTCP are extensive since on one hand the format of the RTCP receiver reports and on the other hand the timing for the transmission of these must be changed. The TFRC-specific information should be transported in the application-specific information of an RTCP receiver report and should contain the four fields shown in Figure 7.10. The timing for the receiver reports must be changed in a way that it is based on the RTT information instead of the algorithm proposed in reference [62].

7.3.3.2 Packetizer and Depacketizer

Several profiles for the transport of standardized audio and video formats in RTP exist [63]. So far no profile for the transport of layer encoded video is defined, a result of the lack of a standard for this technique. Experiences with the development of (de-)packetizers for several audio and video formats [159] have shown that building new SH for this purpose is a rather straightforward task. Depending on what layer-encoded video techniques are to be supported, it may be necessary to build more than one SH. Since it was the goal of this work to demonstrate a more general approach which is not specifically designed for one layer-encoded format, a pseudo-layered format is presented in Chapter 8.

To describe the interaction between the SHs in more detail, an example scenario for in-band retransmission is shown in Figure 7.13. Whenever an RTP packet is sent out on the network the payload will be pulled from the packetizer SH. With the report functionality of the SH architecture the RTP sink SH can inform the packetizer SH about the actual transmission rate. With the aid of this information the packetizer can build the two different types of

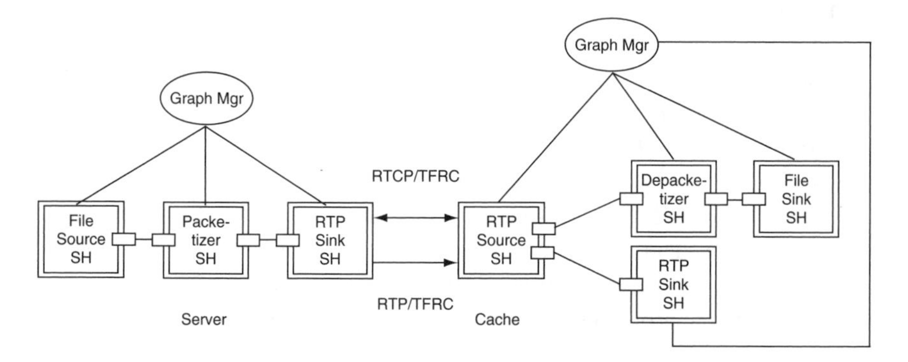

Figure 7.13 Stream handler scenario for in-band FSC.

payloads. For the layer-encoded video the rate information also determines the number of layers that should be transmitted. If the number of layers is known, the packetizer can determine the resulting capacity that is available for the retransmission of segments and determine which type of packet (layer-encoded video or retransmitted segment) will be handed to the RTP sink SH[†] in the case of a pull request from the latter. With each payload the packetizer must also provide the appropriate SSRC information which allows the correct demultiplexing in the RTP source SH at the receiver. Further information about the stream handler architecture is given in Appendix C.

7.4 SUMMARY

In this chapter, a new technique called FSC is presented that allows one to claim a fair share for TCP-friendly sessions for the transmission of layer-encoded video when caches are involved. This technique bears the advantage that on the one hand these sessions actually will get their fair share of the link and on the other hand the quality of already cached video can be improved. In order to prove the applicability of FSC, a simulation environment that consists of three single simulation steps is created. The single steps include: creation of TFRC traces, layer-encoded video transmission, and retransmission scheduling. A series of simulations based on this simulation environment is performed. The results of the simulations state the applicability of FSC, in particular, that in combination with the retransmission scheduling algorithms developed in Chapter 5 a reasonable quality improvement for already cached video can be achieved. It is also shown that another already existing retransmission scheduling algorithm is not well suited for the proposed FSC technique. The simulation results also state the results of the initial simulations made in Chapter 5 where, in contrast to those in this chapter, the initial video objects were created randomly and not based on TFRC traces.

Since the results of the simulations met the expectations, the design for the integration of FSC in an already existing streaming platform is given. An extension to RTP/RTCP is proposed which allows the usage of the TFRC mechanism and, thus, a TCP-friendly streaming between sender and receiver. Additionally, signalling extensions for RTCP are made which allow the demultiplexing of the data that belongs to the actual stream and the data that is retransmitted. For the integration of the FSC mechanism it is only necessary to extend existing protocols; it is not necessary to develop new protocols.

[†] This stream handler acts as a sink in the server's stream handler graph.

8

Scalable TCP-friendly Video Distribution for Heterogeneous Clients

8.1 MOTIVATION

In recent years the variety of different types of clients with access to the Internet has increased. Only a few years ago, the typical Internet client was a standard PC connected via LAN or modem, but today Internet clients are also set-top boxes, PDAs, mobile phones or even game consoles, and the number of wireless clients is increasing rapidly. Their characteristics in terms of computing power, memory space, and access bandwidth vary greatly, thus leading to new challenges for a video streaming architecture.

Layer-encoded video ideally supports such an architecture since it allows an adaptation to link bandwidth, client processing power and buffer size, which is not possible with non-adaptive formats, for example, MPEG-1. A cache can be introduced as a node in a distribution system to address several of the problems associated with this heterogeneity. It separates the long-distance network from the access network and their distinct characteristics concerning throughput, jitter and loss. In this chapter, the focus is on the ability to apply, conditionally, congestion-control mechanisms separately for the server–cache and the cache–client link. For example, it is possible to stream from server to cache with higher bandwidth than from cache to client.

Basically, the goal of the work described in this chapter is to experimentally investigate the applicability of different transport mechanisms in combination with caches, heterogeneous clients and layer-encoded video. The setup for the experiment reflects a typical scenario for an SAS architecture.

In Chapter 7, it was the goal to integrate an optional congestion-control mechanism into the existing streaming system without losing efficiency and

Scalable Video on Demand: Adaptive Internet-based Distribution M. Zink

compatibility. In the following, an extension that allows the streaming of a pseudo-layered format and supports standard clients which have no extended signalling capabilities is presented. The support of standard clients allows a stepwise transition from traditional to SAS-based video distribution. Based on this implementation, several experiments to prove the applicability of the proposed SAS architecture (see Chapter 2) were conducted. First, it is shown how controlled packet dropping can result in a smooth transport of the most relevant layers of a layer-encoded video to a client. Second, the effects of an uncontrolled access network are shown and compared with the controlled approach. Both experiments are performed in a testbed, as well as in the Internet. The experiments also show that a sustainable number of layers can be determined easily by applying the congestion-control mechanism of TFRC [47] on the access link.

8.2 TRANSPORT SCENARIOS

In this section, the different transport scenarios supported by the SAS architecture are introduced. As a result of the ability of server, cache, and client to perform congestion control there are four possible ways different transport mechanisms may be used on the link between server and cache and cache and client. These four possibilities are illustrated in Figure 8.1.

For each of the four scenarios an example is given which shows how a layer-encoded video could possibly be transmitted between server, cache and client. For the sake of simplicity no losses are assumed on both links. Note that only scenarios 1 and 3 allow the usage of standard RTP/RTSP clients. Scenario 1 allows the usage of standard RTP/RTSP server, cache and client and represents commercial systems that are available today. In all other cases (2–4) at least one of the components has to be extended as described in the following. Scenario 2 shows an example in which data between server and cache is streamed uncontrolled while congestion-controlled streaming between cache and client is performed. This scenario allows the usage of an unmodified server while cache and client must be able to perform congestion-controlled streaming. In scenario 3, an example is given in which the client signals the maximum rate it can receive data via RTSP to the cache. Since this maximum bandwidth only allows the transmission of up to two layers, segments from layer 3 are not forwarded from the cache to the client. Nevertheless, the video is streamed from the server to the cache with a higher rate in order to increase the quality of the cached video object. This approach is appropriate in cases where cache storage is sufficient and the probability

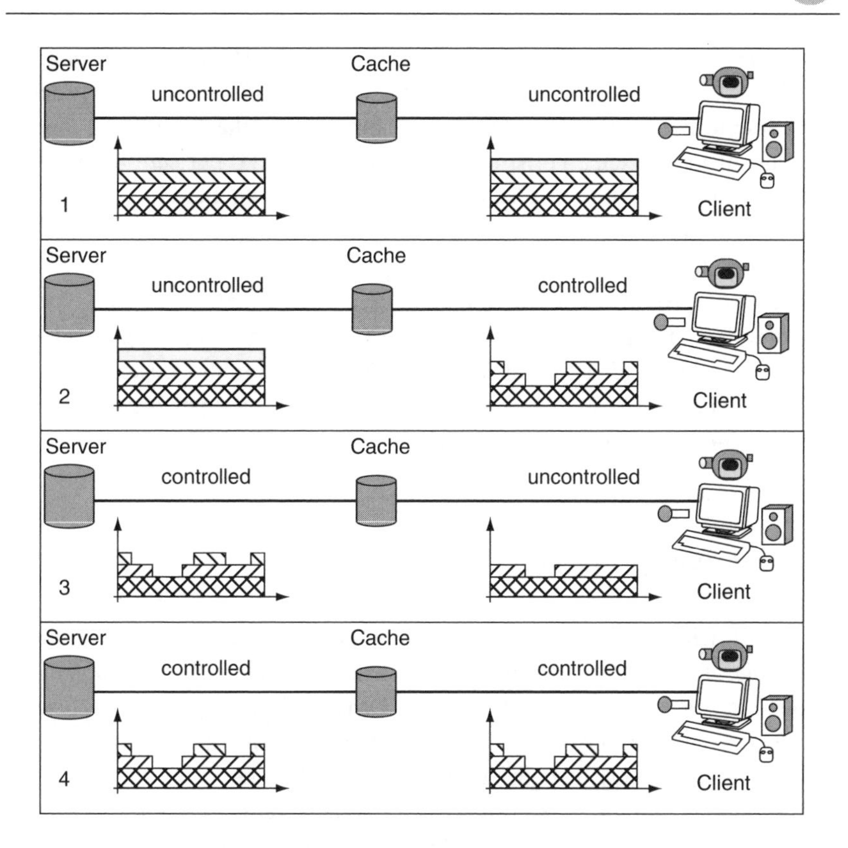

Figure 8.1 Combination of transport mechanisms.

that other clients request this object in a better quality is high. The signalling of the maximum transmission rate by the client is necessary, because no congestion control is performed between cache and client. This can prevent the cache from sending with a rate that cannot be consumed by the client.

In scenario 4, both links are congestion-controlled, so sending signalling information about the maximum rate from the client to the cache is not necessary.

8.3 SCALABLE STREAMING IMPLEMENTATIONS

Owing to the limited information about commercial products [56, 54, 160] the focus is on implementations created by the research community.

Rejaie and Kangasharju [118] mainly focus on the design and implementation of a cache and the goal of adaptively adjusting the quality of a cached stream based on popularity and available bandwidth. The main difference to the SAS approach is the fact that clients in their architecture always have to be able to perform congestion control. In addition, the congestion control mechanism in reference [118] is not integrated in RTP but set on top of it. Another implementation of a TCP-friendly, partially reliable video streaming approach is presented by Feamster *et al.* [98]. No caches are envisaged in this architecture.

Race *et al.* [161] present the implementation of a RAM-based video cache which is designed for the caching of MPEG-2 streams. The usage of RAM instead of a hard disk circumvents a bottleneck on the disk's channel. DSM-CC is used for stream control and the streaming is performed in non-congestion-controlled manner via UDP.

8.4 IMPLEMENTATION

In the following, the design and implementation of the SAS architecture that is capable of supporting heterogeneous clients are presented. The implementation described here is based on the KOMSSYS [162] streaming system which was built to perform research on wide-area distribution systems. In a basic version [159], it consists of server, cache and client and is based on the standard protocols RTP/RTCP, RTSP and SDP. To verify new ideas for scalable distribution systems it has been extended by new functionality and used for several experiments. LC-RTP ([127] and Appendix A) and *gleaning* [4] are two examples of the new functionality that was integrated into KOMSSYS (see Appendix C).

In the remainder of this section, the modifications and new functionality for the data-path-related and control-path-related parts of the streaming system are presented. Finally, it is shown how all of the new pieces are orchestrated in the cache to offer the new functionality.

8.4.1 Data Path

8.4.1.1 TCP-friendly Congestion-Control Mechanisms

In Chapter 7, it was explained why TCP-friendly rate control (TFRC) is well suited as congestion-control algorithm for streaming in the SAS architecture. It is also shown how the integration of TFRC in RTP can result in a congestion-controlled streaming mechanism.

8.4.1.2 Layer-encoded Video Format (Layer Dummy)

The *layer dummy* (LD) format that is used for the experiments described section 8.5 is presented in this section. Afterwards, the modifications that have to be made to perform lossless transmission into caches using layer-encoded video are shown. The decision to use the LD format is made because it should first be investigated whether the modifications on the data and control path proposed here meet the expectations. In addition, the LD format makes measurements easier, since the specific RTP payload format allows an extensive logging (see section 8.4.2). The way LD is designed allows an easy integration of layered formats such as SPEG [81]. Thus, the LD format is designed with properties similar to SPEG. It is assumed that the format is hierarchically coded, i.e., a segment of higher layer data is worthless if the corresponding lower layer data segment has been lost. It is also presumed that the bitrate is constant, and that all layers have equal segment sizes and equal bandwidths. The RTP payload for this format includes a header as shown at the bottom of Figure 8.2, which includes a sequence number and a layer field. The latter specifies the layer of the video data that follows the payload header. The MTU size chosen by RTP can differ from the segment size; therefore, the payload of one RTP packet can contain several segments.

For the experiments (see section 8.5) a four-layer format as shown at the top of Figure 8.2 is used, where the first digit specifies the sequence number and the second the layer of the segment.

layer dummy format

3,4	7,4	11,4	15,4
2,3	6,3	10,3	14,3
1,2	5,2	9,2	13,2
0,1	4,1	8,1	12,1

sequence number layer

RTP packet with layer dummy payload

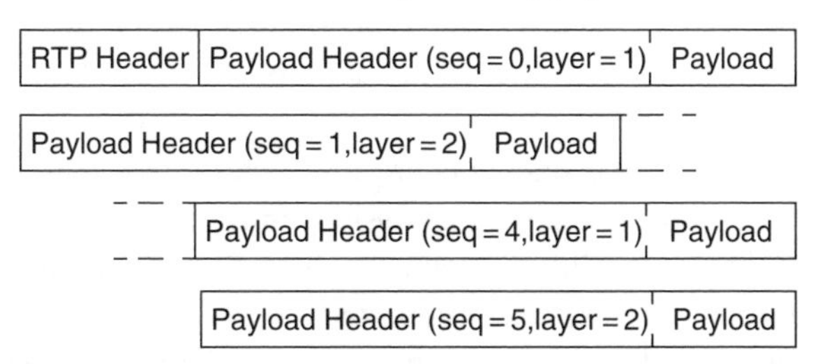

Figure 8.2 Layer dummy format and RTP packet.

8.4.1.3 LC-RTP Extensions

In preceding work (see Appendix A), an extension to RTP that provides lossless transmission of A/V content into caches and, concurrently, lossy real-time delivery to clients is proposed. Here, an extension to LC-RTP is made such that two different types of transmission of missing segments are possible. Depending on different factors such as popularity of the video and kind of client it may only be necessary to transmit the losses that occurred during the transmission or in addition transmit the segments of the layer-encoded video that were not transmitted at all (for example, in the case of congestion). For the second case no modifications need be made to LC-RTP, since all missing segments are transmitted as with the original LC-RTP. In the first case, two modifications to LC-RTP must be made:

- The sender stores a list of the segments that are actually sent into the network. With the aid of this list the server can identify which segments have to be transmitted and which do not. The client sends requests for the transmission of missing segments until all segments from its list of missing segments are received, a maximum number of retries is reached, or a BYE message from the server is received.
- When the server notices that the client request contains only unsent segments it sends the client a BYE message to stop it from sending further requests. Since the client is already in the loss collection phase it interprets this message differently from a BYE that is sent at the end of the initial transmission and stops sending requests for retransmission.

This extension to LC-RTP requires only minor changes to enhance its functionality and no new protocol messages need be introduced. In order to distinguish between the two retransmission methods the *subtype* field of the application-defined RTCP packet is subdivided as follows. The first bit is used to indicate which retransmission method should be applied ($0 =$ only losses, $1 =$ losses and not initially sent segments) and the remaining four bits are used to identify the type of the packet. In Figure 8.3, an example transmission from server to cache is shown. Owing to bandwidth limitations only three out of four layers are sent to the cache. During the transmission some of the segments are lost. These losses are recognized by the cache and requested for retransmission. Since the cache is not aware of the fact that the server did not send the segments belonging to the fourth layer, those segments are also requested. The server keeps track of the segments that have been sent to the client initially and retransmits only those segments. If the client requests only packets that have not been sent,

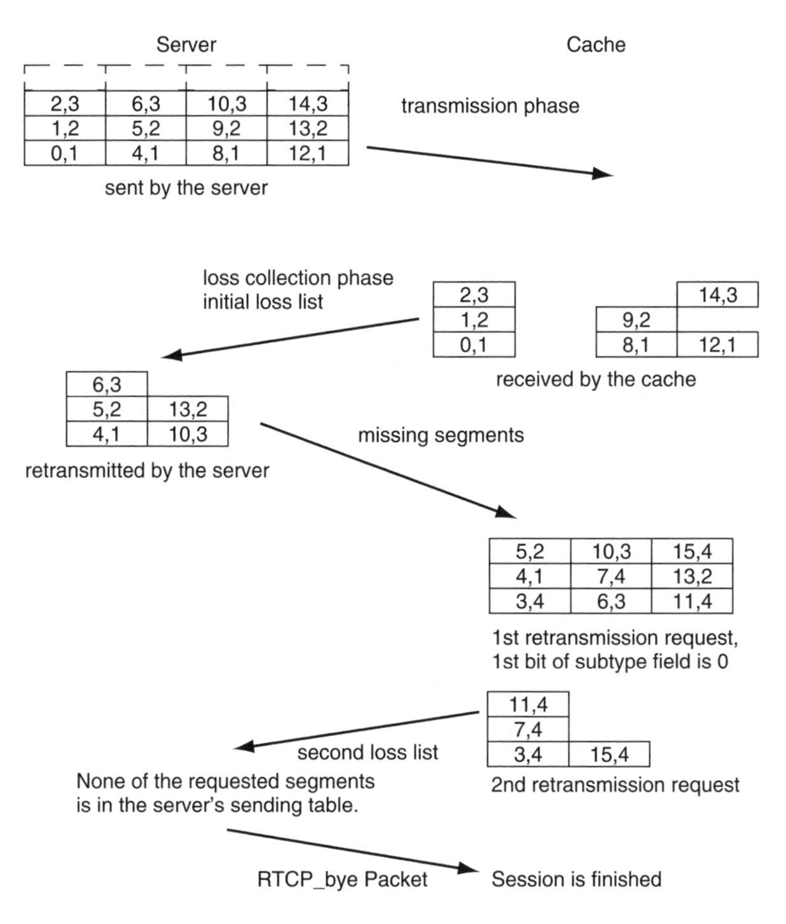

Figure 8.3 LC-RTP for layer-encoded video.

the server sends a RTCP BYE message and the retransmission phase is terminated.

8.4.2 Signalling

In the KOMSSYS streaming environment, RTSP is used as application signalling protocol. Here, it is shown how RTSP can be used to allow members of a streaming session to negotiate whether they are capable of a congestion-controlled streaming session or not. To achieve this capability a new *transport-protocol* identifier in the *Transport* header, called *'RTP/TFRC/UDP'*, is introduced. This would, for example, allow a non-RTP/TFRC-capable client to send a *Setup* message as shown in Figure 8.4

and, therefore, initiate a non-congestion-controlled session between itself and the cache. The cache on the other hand can modify the *transport-protocol* identifier and initiate a congestion controlled session between itself and the server.

If the cache is capable of establishing a congestion-controlled session towards the server and the client, the tag stays unchanged and is forwarded to the server. An equally extended server replies with the appropriate *transport-protocol* identifier, depending on whether it is RTP/TFRC capable or not. The cache modifies the identifier according to its own capabilities and the client's initial request and forwards the message towards the client. A complete communication between server, cache and client regarding the RTSP *Setup* message is shown in Figure 8.4.

Since a cache should also support clients that are not capable of congestion control (see section 8.1), it can automatically insert the *transport-protocol* identifier in the *Setup* message that is forwarded to the server and remove it from the server's reply before it is forwarded towards the client, thus allowing a congestion-controlled stream between server and cache and a standard RTP/UDP stream between cache and client.

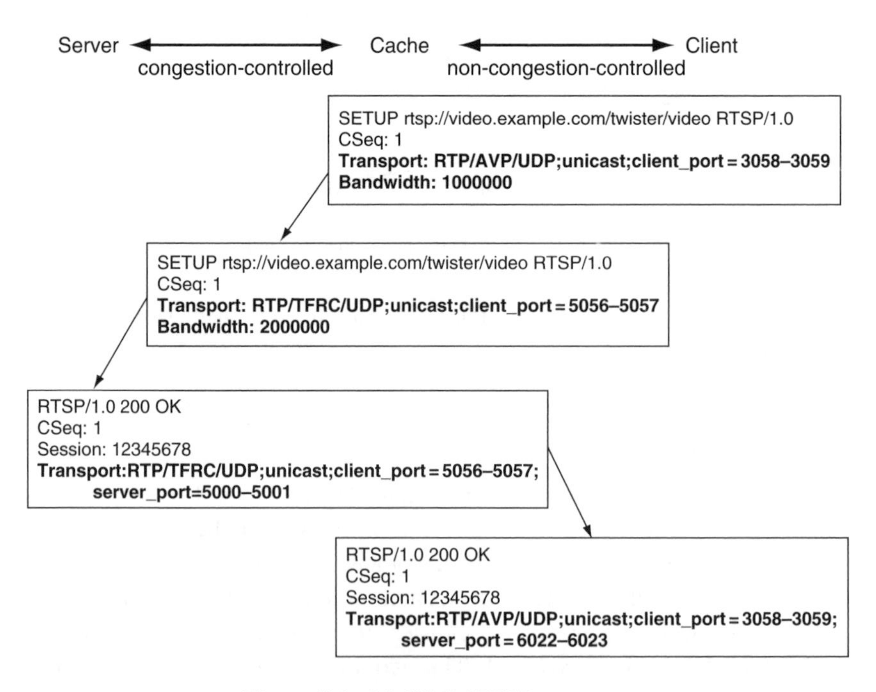

Figure 8.4 Modified *SETUP* messages.

Additional information that is signalled by RTSP is the maximum bandwidth at which a stream can be sent to the client or the cache. For this purpose, RTSP provides the *Bandwidth* request header field, that describes the estimated bandwidth available to the client, expressed as a positive integer and measured in bits per second [16]. In the case of the KOMSSYS implementation this header field is also added to the *Setup* header and used by either the server or the cache to limit the maximum bandwidth TFRC used to stream the content. In section 8.2, the functionality that allows a faster streaming than the default rate between the server and the cache and how the cache serves as a buffer for the client are described. To achieve this functionality, the information in the *Bandwidth* request header field is modified by the cache. The *Setup* signalling for this case is also shown in Figure 8.4.

8.4.3 Putting the Pieces Together: Cache

In Appendix C, the stream handler architecture that builds the basis for the KOMSSYS streaming platform is presented. Stream handlers are media processing units which can be bound together by a controlling entity into a directed graph. This approach allows the creation of new functionality in either server, cache or client by reusing existing stream handlers.

Figure 8.5 depicts the streaming graph at the cache including both alternatives for congestion-controlled and standard RTP/UDP-based streaming towards the client. Conditional write-through caching is enabled by the use of the PacketMultiplierSH that creates a copy of each received segment, which is then stored at the local disk. This part of the streaming path is only activated if a positive decision to store the video in the cache is made by the caching strategy. The two alternative paths that handle the data forwarding towards the client are created on the basis of the RTSP information that is received from the client. If the client RTSP *Setup* message contains the RTP/TFRC *transport-protocol* identifier, the upper path in Figure 8.5 is created; if not (in the case of RTP/AVP) the lower path is created. The PushPullSH in the upper path implements a queue with limited length that allows controlled dropping. In the lower path, all data is forwarded as it arrives at the cache.

Both paths have additional functionality that is only needed for the measurement in the following experiments. If logging is requested, a log entry is generated in the respective trace files for each layer of a frame that is received either at the cache or the client.

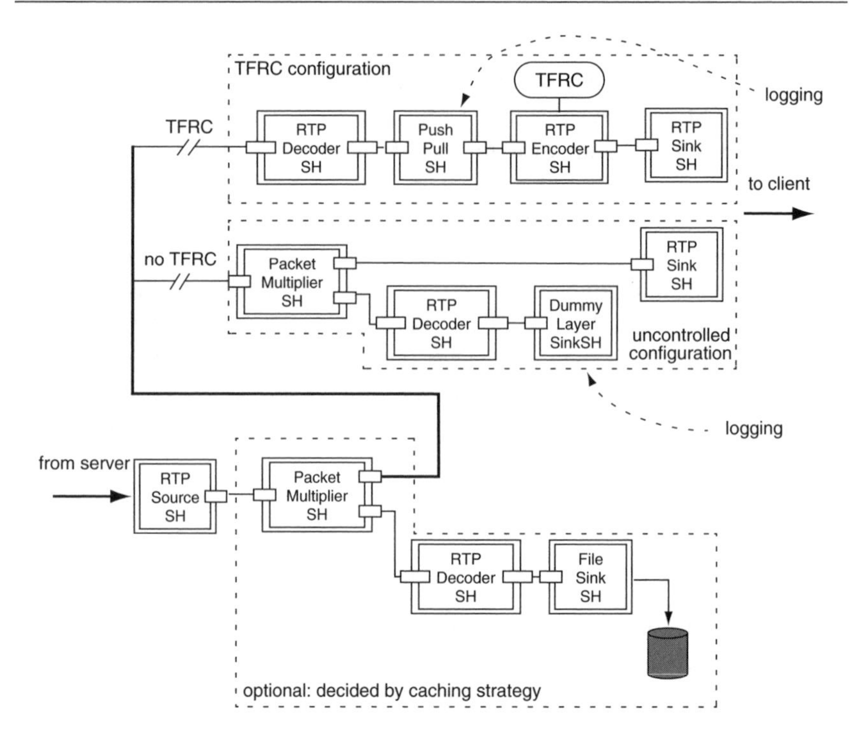

Figure 8.5 Cache configuration for TFRC downlink and for uncontrolled downlink.

8.5 EXPERIMENTS

8.5.1 Setup

The goal of the experiments is to verify that the functionality described in section 8.2 can be realized by an implementation. Therefore, the KOMSSYS streaming platform is extended by the functionality described in section 8.4. Two sets of experiments were conducted with the modified platform. The first one is a testbed-based experiment. The five computers (Server-NISTNet1-Cache-NISTNet2-Client) used for this experiment were standard Pentium-III PCs (850 MHz) with 256 MB of RAM and Linux 2.4 as operating system. During the experiment, KOMSSYS server, client and cache were used. The use of two additional NISTNet [163] network emulators allowed the reduction of the link conditions between server and cache and between cache and client from the original capacity of a switched 10 Mbit/s Ethernet. The measurements described in the following section were executed in an environment without additional network traffic, since the testbed is not connected to any other network. For the experiments a bandwidth of 1 Mbit/s for the server–cache link, and 512 kbit/s for the cache–client link is chosen. This allows full quality streaming between server and proxy, while the quality must be reduced on the link between cache and client owing to a limited bandwidth. The bandwidth of the layer dummy video at full bit rate is 1 Mbit/s, with each layer making up an equal share of 256 kbit/s. The testbed setup is shown in Figure 8.6

To complement the artificial network setup with real world data, a measurement over the Internet between Oslo/Norway and Darmstadt/Germany was performed in addition. The setup for this measurement is shown in Figure 8.7. The test machines are different Pentium-III Linux 2.4 machines. The network was fairly unloaded during the test.

8.5.2 Measurements

In the following, measurement results from both test setups are presented. The results are shown in graphs that depict the number of received layers at the cache and at the client. With the logging mechanisms presented in section 8.4.2 the number of layers that are either stored at the cache or received at the client is measured. In the case of hierarchically layered codecs, the loss of a segment with a lower layer number implies that all arriving segments with higher layer number, which belong to the missing segment, have to be considered lost as well. Under this consideration, one graph that shows the valid, arrived layers for every individual frame in the video is created. The frequency of layer changes makes it hard to visualize the average number of layers that are received. To provide a better insight, a second graph is shown, which presents the same quality in terms of layers once for each frame in the movie, but instead of just showing the number

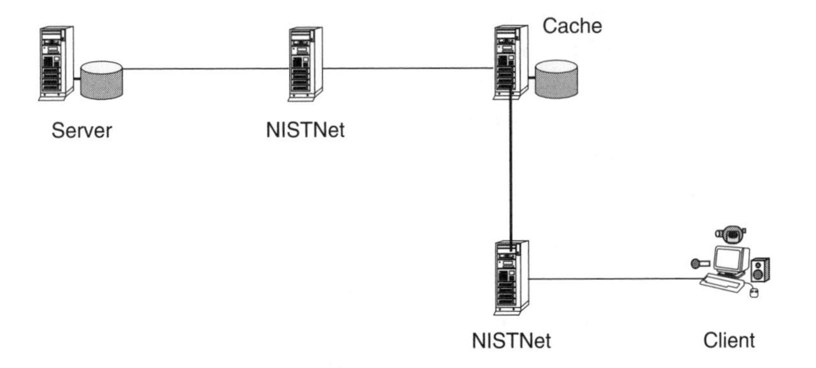

Figure 8.6 Tested setup.

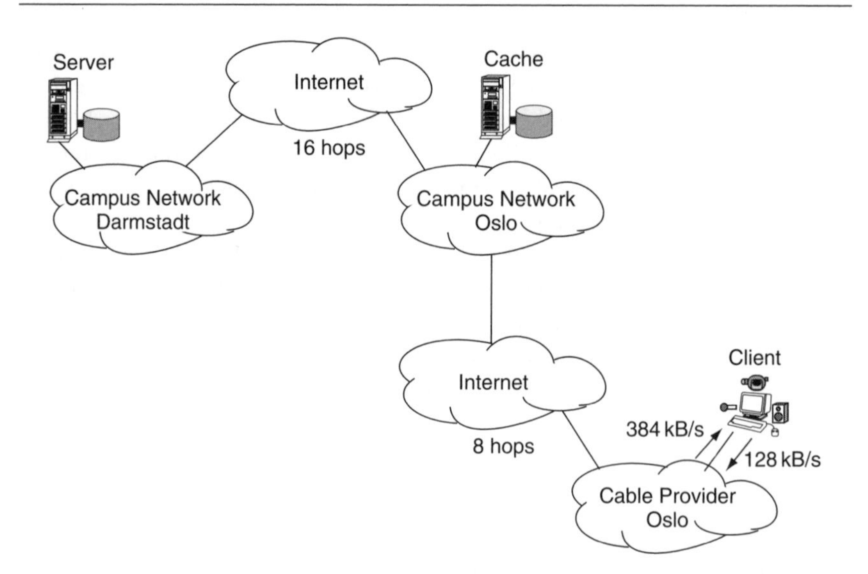

Figure 8.7 Internet-based measurement.

of layers for the individual frame, the average of the frame and the previous 100 frames is shown. This presentation hides short-term quality changes, but makes it easier to identify mid- and long-term changes in the quality development.

8.5.2.1 TFRC in the Access Network

In this experiment, the client demands less bandwidth from the cache than the cache demands from the server. This allows the cache to be filled with a higher-quality version of the video object, and clients that request the same object later in time are able to receive higher qualities as well. In Figure 8.8, the testbed-based results for this experiment are presented.

The detailed observation of layers received at the client shows a very low loss rate for packets that contain layer 1 and layer 2 data. This implies that TFRC chooses a sustainable bitrate, which can in this case support a load of two layers. Additional packets of layer 3 are inserted into the stream for bandwidth probing, and have a considerable probability of getting lost.

The use of TFRC allows the usage of this additional bandwidth above the sustained bandwidth in any application-defined way. It is inappropriate to use it for faster-than-real-time transmission of the base layer because the high loss probability would make the additional use of retransmission necessary. For a hierarchically layer-encoded video, however, this additional layer, which is too unstable to be used for transmitting another layer, may be used to implement retransmission (as proposed for fair share claiming in Chapter 7) or forward error correction for the base layer.

By examining all scenarios of TFRC use, a strong instability in the early phase of the connection is observed, compared to the stable operation when the appropriate bandwidth has been found. Since the duration of this start phase is relatively long, it seems reasonable to enter the initial slowstart phase with the bandwidth of a previous connection rather than a full slow-start as proposed in [164].

8.5.2.2 UDP in the Access Network

In a second measurement, the effects of using UDP in the access network combined with TFRC in the backbone is investigated. This investigation is legitimate for several reasons. First, bandwidth in access networks that use DSL or cable modems is frequently restricted by shaping but provides a guaranteed minimal throughput in the provider's network. If caches are deployed at a provider's site, the customers should be able to exploit this guaranteed bandwidth in such a way that an additional consideration of TCP friendliness in the access network is not required. Second, standard-compliant clients do not use TFRC. Thus, it is important to understand the interplay of using TFRC in the backbone and regular RTP/UDP in the access network. The results of the experiments show that when the access link becomes the bottleneck link, random loss occurs in the case of UDP. This should be compared to the controlled loss that is possible when the available bandwidth can be estimated based on TFRC information. It becomes apparent from the detailed graphs in Figures 8.8 and 8.9 that the UDP approach suffers from one problem. The rate of valid packets is not sustained, which results in frequent quality changes. Since the access link can support approximately two layers and the cache forwards approximately four layers, the probability of a successful arrival of a layer 1 block is 0.5 and a layer 1 and a layer 2 block 0.33. This yields an estimated average number of 0.66 valid layers. The results in Figure 8.9 support this estimation. The difference in TFRC performance between the testbed-based and the cable-based access network becomes visible by comparing the testbed results with the real world traces in Figure 8.10. Whereas the congestion-controlled approach yields about two-thirds of the uncontrolled throughput on the access link, it is just one-third in the real-world scenario. This limited throughput is probably due to artificial limitations of the TCP throughput to 384 kbit/s, which are defined in the service agreement. However, it can be observed in both scenarios

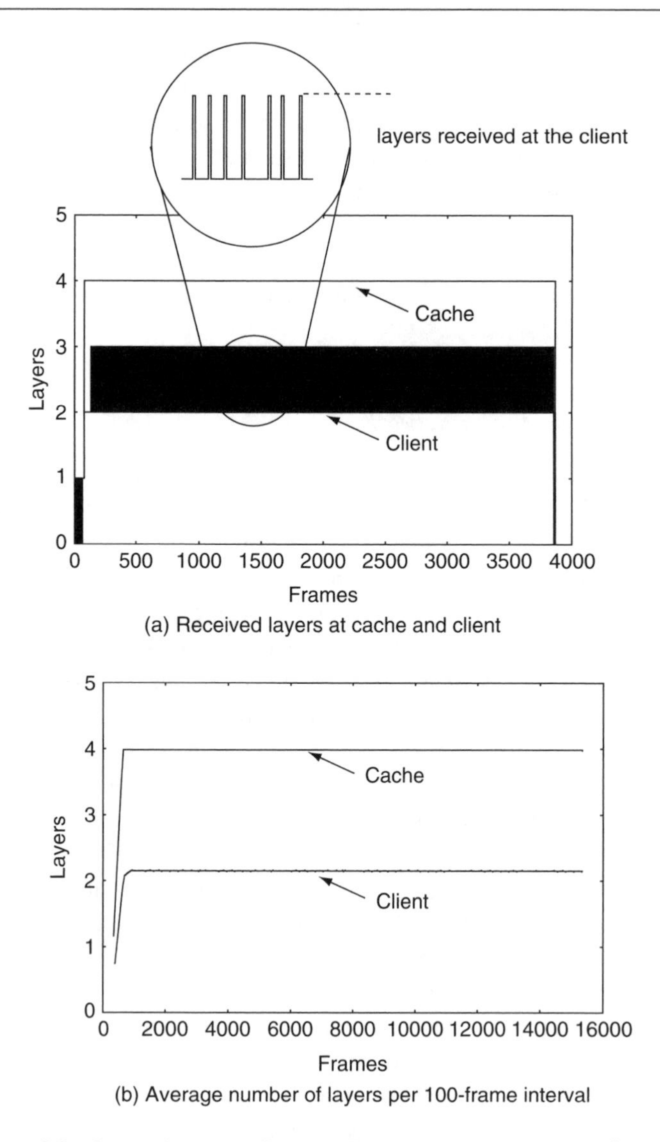

Figure 8.8 Congestion controlled by TFRC in the access network (testbed).

that the short-term quality changes with congestion control are considerably lower than without. In the observed case, it does not seem viable to use the bandwidth available between the server and the cache for a complete transfer of movies into the cache, because of the contractual limitations of the downlink. However, since a bandwidth very close to the demanded

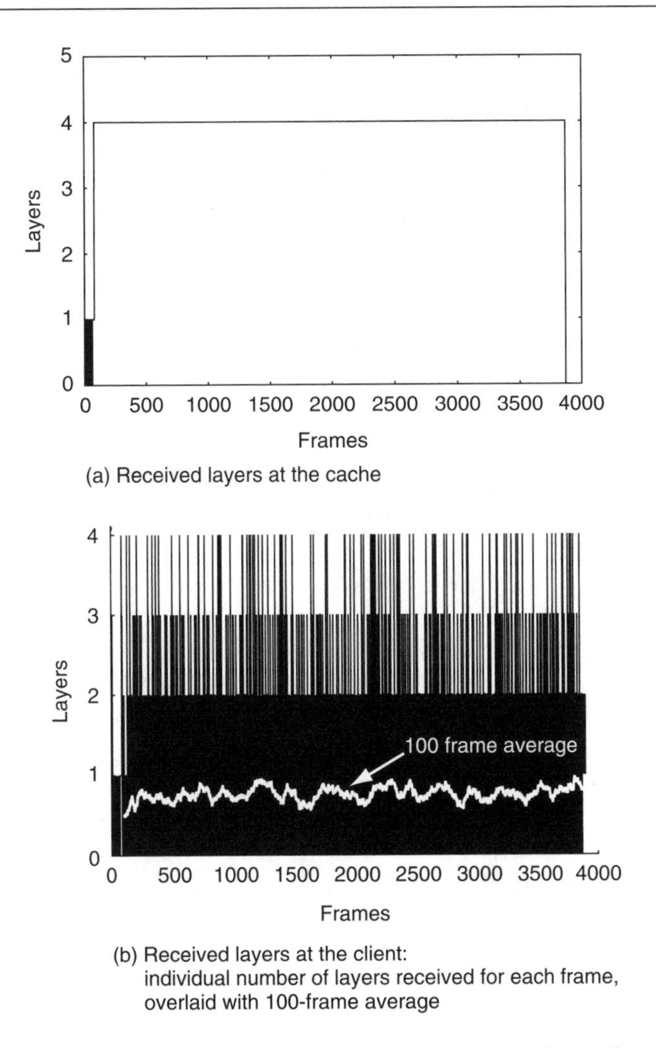

Figure 8.9 Uncontrolled UDP in the access network (testbed).

bandwidth is made available, it is appropriate to allocate it for faster-than-real-time transmission and retransmission.

The incursion of the bandwidth can only be explained by some unexpected behaviour on the access network in Oslo.

8.5.2.3 Hierarchical and Non-hierarchical Codecs

Since the packet droppings are entirely random, it is important to notice the disadvantage of a hierarchical layer-encoded video codec: a codec that uses

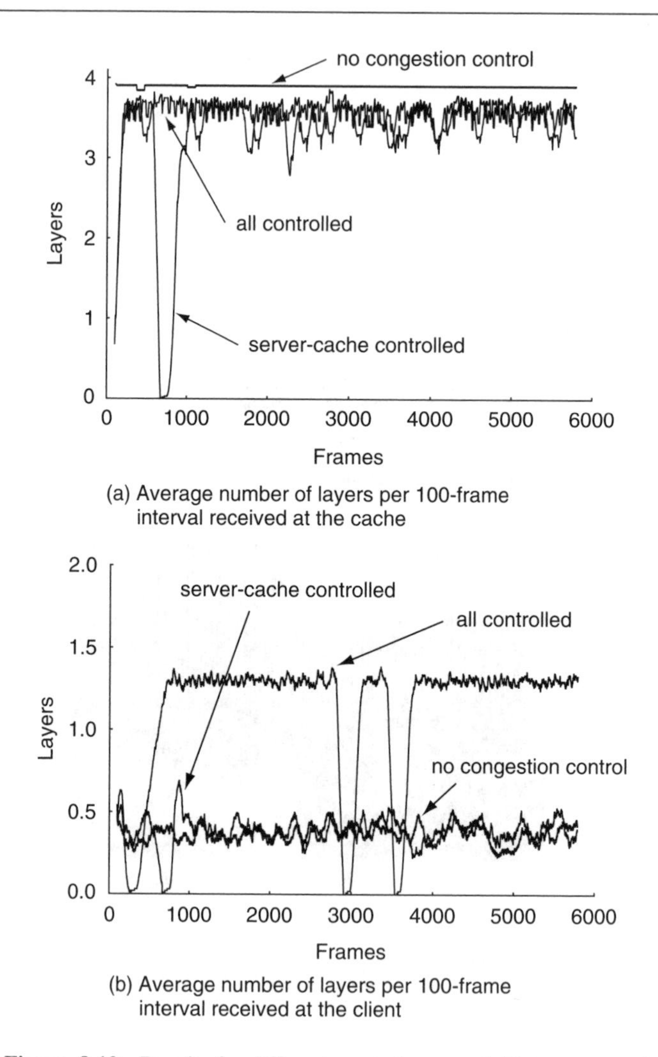

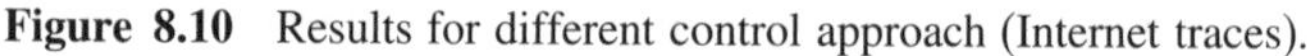

Figure 8.10 Results for different control approach (Internet traces).

independent layers (as described in section 3.3.3) would have considerably more stable results. A comparison with the results that a non-hierarchical codec would yield is presented in Figure 8.11. The two graphs in the figure show two alternative interpretations of the same trace files, a hierarchical interpretation on the left and a non-hierarchical interpretation on the right. Since no filtering takes place in the TFRC case unless the forwarding queue in the cache is full, there is no need to consider the preference of forwarding packets with lower layer number. Also, in the non-hierarchical case, the number of forwarded packets would be identical even if the queue did

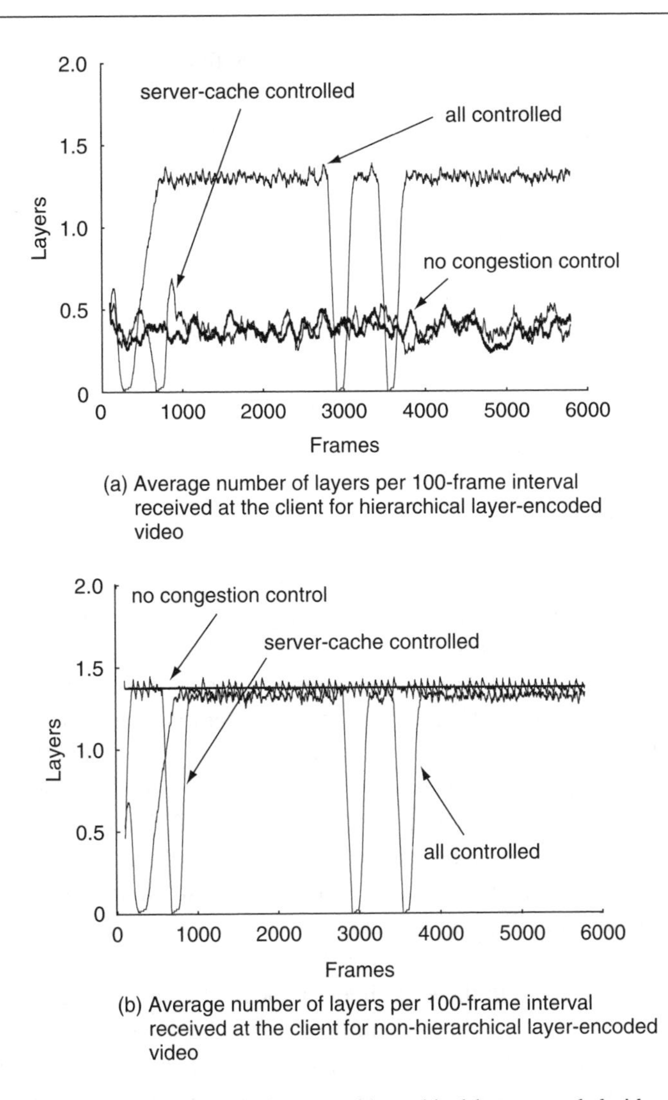

Figure 8.11 Hierarchical vs. non-hierarchical layer-encoded video.

not prefer selected packets. Obviously, the congestion-controlled approach behaves identical for both kinds of encoding. Both cases that are uncontrolled in the access network achieve the same throughput in this case, which is very close to the connection's limit. Nevertheless, the completely uncontrolled case allows for a more stable bandwidth in this case, i.e., fewer quality changes occur than with the congestion-controlled approach. The most surprising observation in the non-hierarchical graphs is certainly that this smoothness is not achieved for the approach that uses congestion control in the backbone and no congestion control in the access network. Although the number of packets that are forwarded onto the access link in an uncontrolled manner far exceeds the available bandwidth, the rate fluctuations of the controlled approach are disseminated to the access network. This observation leads to the conclusion that even if simple shaping is performed in a cache that separates the backbone from the access network, a queue should be introduced into the forwarding path to hide the bandwidth fluctuations.

The results of this investigation show that, especially in the case of an uncontrolled link between the cache and the client, the use of MDC would be beneficial to hierarchical layer-encoded formats.

8.6 SUMMARY

In this chapter, the architecture and implementation of a scalable, TCP-friendly video distribution system for heterogeneous clients is presented. This architecture allows the usage of clients with and without congestion control while a congestion-controlled transmission is always performed between server and cache. Congestion control in the system is achieved by the integration of TFRC in RTP. After the modifications in RTP, an extension to LC-RTP to support the lossless distribution of layer-encoded video into the cache is presented. In addition, it is shown how RTSP signalling can be used to negotiate the type of streaming (congestion-controlled or not) and maximum bandwidth between the single entities of the streaming system. Based on these implementations experiments in both a testbed and the Internet are conducted.

The results obtained in the experiment indicate that all connections should be congestion-controlled, but certainly if hierarchical layer-encoded video is used. With standard clients that cannot perform congestion control on the access link, the usage of non-hierarchical layer-encoded video would be an option. It must be mentioned that in the non-hierarchical case the uncontrolled approach performs best, but is not a TCP-friendly approach.

In addition, this investigation shows that there are cases in which the type of layer-encoded format (hierarchical vs. non-hierarchical) can play an important role in the performance of a video distribution system.

9

Improved Video Distribution in Today's Internet

9.1 IMPROVEMENTS THROUGH SCALABLE ADAPTIVE STREAMING

Throughout the course of this book it has been shown that two-dimensional scalable streaming in today's Internet is possible. A new architecture for scalable adaptive streaming (SAS) is presented which combines system and content scalability. The SAS architecture consists of mechanisms which allow the delivery of layer-encoded video in an acceptable quality to the user while the overall amount of network traffic is reduced and, thus, the overall scalability of the system is increased. The new mechanisms are based on Internet technology that is available today. The applicability of these mechanisms is shown in the preceding chapters.

One major issue is the introduction of quality variations by the combination of adaptive (due to network congestion) streaming and scalable content. These variations are a natural side-effect of the adaptive streaming mechanisms which should be kept to a minimum in order to keep the quality of the transmitted stream acceptable. Since caches are involved in the distribution of the video objects, additional mechanisms can be applied to reduce the variations of a layer-encoded video.

Owing to the lack of a subjective assessment of variations in layer-encoded video, part of the book is dedicated to this issue. This investigation is the first of its kind and results in a set of guidelines being used to develop transport mechanisms in SAS and an objective quality measure. The results from the subjective assessment influenced the development of a new objective measure, the spectrum, that is more suitable than the well-known peak signal-to-noise ratio (PSNR).

Scalable Video on Demand: Adaptive Internet-based Distribution M. Zink

To improve the quality of layer-encoded video, either stored on a cache or streamed through a cache to the client, new mechanisms to reduce layer variations are developed. It is shown that an optimal solution is computationally infeasible and, thus, new heuristics are proposed. The development of these heuristics is also influenced by the insights gained from the subjective assessment. A simulative investigation shows the performance and applicability of the new heuristics. To compare the heuristics with each other, the spectrum is used as a quality metric. In addition to the heuristics, different retransmission scheduling approaches are introduced which maximize either the quality of the cached layer-encoded video or the one that is streamed to the client. The results of the simulations show that retransmission scheduling is an effective means of reducing variations in layer-encoded video. With the fair share claiming (FSC) technique, a new transport mechanism is presented which allows the retransmission of missing segments from the server to the cache.

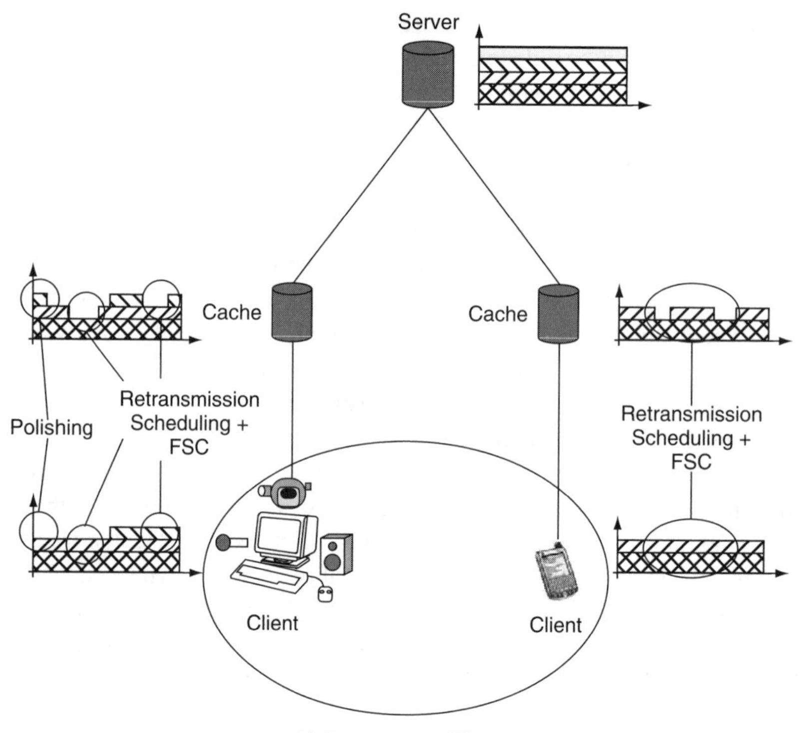

Figure 9.1 Improved Internet-based video distribution.

To support heterogeneous clients in a video distribution system extensions for signalling and data transport are made. These extensions are implemented in an existing experimental streaming platform. On the basis of this implementation measurements in both a testbed and the Internet are performed to prove the applicability of the proposed protocol extensions. The results of these measurements clearly show the benefits of congestion-controlled streaming in combination with layer-encoded video.

Finally, an investigation into the reduction of variations on layer-encoded video in the case where retransmissions cannot be performed resulted in the polishing technique. This technique was further developed into a fine-grained cache replacement algorithm that can perform replacement for layer-encoded video based on the popularity of the video objects.

Taken together, the new mechanisms presented in this book make it possible to construct VoD systems in today's Internet that outperform existing ones. The results of the performed investigations can be combined to build a new video distribution system; or existing systems can be extended by one or more mechanisms presented in this book. Figure 9.1 gives a general overview of how the mechanisms presented in this book improve video streaming in the Internet.

9.2 OUTLOOK

Applying the results presented in this book can lead to more efficient and better quality streaming of video data in today's Internet. The scalability of the distribution infrastructure is increased by the usage of caches on the one hand and layer-encoded video on the other. The combination of both scalabilities, system and content, not only increases the quality of the received stream at the client but also allows the support of heterogeneous clients. Thus, using these new mechanisms, video-on-demand systems can be created that are more scalable, offer a better quality, and support a large variety of clients.

The presented approaches can either be used in combination or stand alone. For example, the results of the subjective assessment could also be valuable for researchers who investigate objective quality metrics for layer-encoded video. The retransmission scheduling and polishing mechanisms can be integrated in already existing systems. In particular, the polishing-based cache replacement method can be applied in caches fairly easily since no other parts of the cache, such as the communication protocols are affected.

In the future it may be necessary to adapt the presented transport mechanisms to new evolving scalable encoding formats. One example is the case of multiple description coded video (MDC). The results of the measurement performed in Chapter 8 indicate that the usage of non-hierarchical encoding formats (such as MDC) can be advantageous in comparison to layer-encoded video in certain cases. Unfortunately, this task is not simply executed by exchanging layer-encoded video through MDC video. Rather it starts with a new subjective assessment for MDC video to verify that the assumptions that retransmission scheduling and polishing are based on are also valid for this new encoding format.

Another interesting research area is the realization of streaming in a peer-to-peer environment in combination with scalable video formats. Some of the existing problems were identified in [165] and it is shown that MDC is a well suited encoding format for streaming in peer-to-peer networks, but there are still many open issues to solve.

Some of the mechanisms presented in this book provide solutions for the support of heterogeneous clients, which also includes wireless clients. Yet, since the characteristics of the wireless channel are often different from wired channels, new or modified transport mechanisms are necessary. Initial work [166] has shown that the modification of TFRC can improve the quality of a stream in an UMTS network. Further investigations are necessary to show whether caches can provide additional support for wireless clients. For example, a cache could also act as a gateway which is located on the transition between wired and wireless network and allow a differentiation between congestion-based and fading-based losses. New research issues in this area will also depend on the technologies fourth generation wireless networks will be based on.

Very important issues in relation to video-on-demand are digital rights management (DRM) and copyright protection. There is a high demand for such mechanisms from the content providers, but none are available for a SAS architecture so far.

Appendix A: LC-RTP (Loss Collection RTP)

A.1 MOTIVATION

In this appendix a protocol is presented which allows the lossless transport of audio and video data into caches while, in parallel, this data is streamed to one or several clients. In contrast to existing approaches on reliable multicast (see section 3.6) the protocol presented here was especially designed for use in content distribution infrastructures for multimedia data.

An important characteristic of A/V content is the fact that it should be transmitted in real-time. This implies that a client cannot wait too long (assuming a limited buffer) for any resent packets instead of displaying the current data, so the normal data flow must persist and any retransmission must happen aside from the normal data flow. The situation is different for caches where the arrival of data is not time-critical (at least not as critical as at the client). Therefore, additional functionality in server and caches in combination with LC-RTP allows a lossless transport into caches while data is streamed to clients as usual.

LC-RTP is designed as an RFC-compliant extension to RTP for reliable file transfer that requires no infrastructure modifications except on servers and caches. LC-RTP provides lossless transfer of real-time data by using loss collection (LC). The sender sends RTP packets via multicast to all receivers (clients and cache servers) in the multicast group. If a cache server detects a packet loss during the transmission it will be memorized in a list. At the end of the session, caches which are storing the video from this multicast transmission request the missing parts from the sender. The sender retransmits all missing blocks and waits until no more packets are requested. LC-RTP works also in simple unicast mode where data is streamed from the server through the cache to the client.

Scalable Video on Demand: Adaptive Internet-based Distribution M. Zink

In section A.2 the standard protocol set for A/V streaming in the Internet and the loss-collection extensions are presented. The design and the general functionality of LC-RTP are presented in section A.3 while section A.4 shows the extension of standard RTP to support LC-RTP. These protocol extensions were implemented in the KOMSSYS streaming platform (see Appendix C) and measurements based on this implementation are presented in section A.5.

A.2 PROTOCOL SET FOR STREAMING MEDIA

In the Internet, one set of protocols is currently adopted – partially or completely – by companies in their products for streaming media (Apple, Real Networks, SUN, IBM, Cisco, FVC.com, etc.). These protocols are the combination of RTSP/SDP for stream control and RTP/RTCP for streaming.

A.2.1 RTSP/SDP

The Real-time Streaming Protocol (RTSP, [16]) is an IETF RFC that is supposed to be used in conjunction with various other protocols. Its functionality is not generic but rather concentrated on stream control. It references elements of HTTP to which it is weakly related. It can be used with either TCP or UDP as an underlying transport protocol. The data transfer protocol that is mentioned in the RFC and that interacts most closely with RTSP, is the Real-time Transfer Protocol (RTP, [62]). The same approach applies for the session description protocols; although no fixed session protocol is defined, the RFC specifies the interaction with the Session Description Protocol (SDP, [65]).

SDP is originally considered as a companion protocol for SAP, the Session Announcement Protocol. However, besides this mode of distribution for session information, others like download from the web or email distribution are also compatible with this kind of information.

A.2.2 RTP/RTCP

RTP (Real-time Transport Protocol) was created to transport real-time data over the Internet. VoD, Internet telephony, MBone-conferences, and all video- and audio-conferences impose specific time restrictions on how the data is delivered. RTP provides payload type identification, sequence numbering, time-stamping and delivery monitoring, and supports multicast if the underlying protocol provides this service.

Table A.1 Protocol set

Reliable file transfer and real-time streaming	
LC-RTP	LC-RTCP
• RTP-compatible until RTCP BYE message	• RTCP-compatible • user application-defined RTCP packets
• use RTP header extensions	
• continuous byte count	• loss-list report receiver to sender
• retransmission after reception of loss lists	• retransmission request after random waiting time
Stream control and sequencing	
RTSP	SDP
• standard protocol • use SDP	• standard protocol • specifies play range • different sources for data segments

Usually RTP is used over UDP, as UDP allows multiplexing and does not have any retransmission schemes such as TCP. RTP is used together with RTCP (RTP Control Protocol [62]) which allows a quality monitoring of the network connection and has minimal control over the session. Furthermore, RTCP can be used to identify the sender. The main task of RTCP is to send periodic control packets to all members of the session using the same distribution mechanisms as the data packets. The resulting protocol set is listed in Table A.1, including the tasks that are handled by each protocol.

A.2.3 LC-RTP

RTP with Loss Collection (LC-RTP) implements a unified protocol for stream transmission that is compatible with RTP, and reliable transfer of content into the cache servers. It solves these problems by making RTP reliable, while the ability is maintained for non LC-RTP capable clients (standard RTP clients) to receive an LC-RTP stream as well. The functionality of LC-RTP is described in section A.3

A.2.4 LC-RTCP

Just as RTP has a companion protocol, RTCP, for the exchange of information about the data transfer, LC-RTP requires a companion protocol

LC-RTCP, which needs to be RTCP-compliant. In application-defined RTCP packets, the receivers inform the sender about their losses after the reception of the BYE packet, unless all of its missing packets have earlier been reported by another receiver.

A.3 LC-RTP DESIGN

In an environment for A/V-caching it is absolutely necessary that the cached version of the content in the cache is stored 100% correctly to avoid error propagation towards the client. With the use of standard RTP on top of UDP, information that gets lost during transmission is also lost to the caches. The problem is that these errors would be transmitted with every stream that is forwarded from the cache server to a client. In any case that should be avoided, since it has to be regarded as a degradation of the service quality. During each transmission data can get lost and, thus, lead to a higher error rate in stored copies.

LC-RTP solves these problems by making RTP reliable, while the ability is maintained for non LC-RTP capable clients (standard RTP clients) to receive an LC-RTP stream as well.

To describe LC-RTP the transmission process is divided into two parts. The first part works almost like a regular RTP transmission and ends after the transmission of the original content followed by the transmission of a BYE message. The second part follows this BYE message and is used to retransmit all lost data. In this scenario, the receiver is a cache that has received a request from a client but that has recognized that the requested content is not stored locally; therefore, a request forwarding to the original or to a cache located upstream towards the server is performed. Figure A.1 gives a general overview of the different steps that are executed during an LC-RTP session.

A.3.1 Unicast vs. Multicast

Depending on the popularity and the availability of IP multicast, it might be required to perform the data distribution via either unicast or multicast. Therefore, it was an additional design goal to support both transmission modes.

The unicast mode is simpler than the multicast mode, since only one cache and one client are served. Also the retransmission phase is simplified because loss reports are only generated by one receiver. Compared to the multicast model the data must be forwarded through the cache towards the

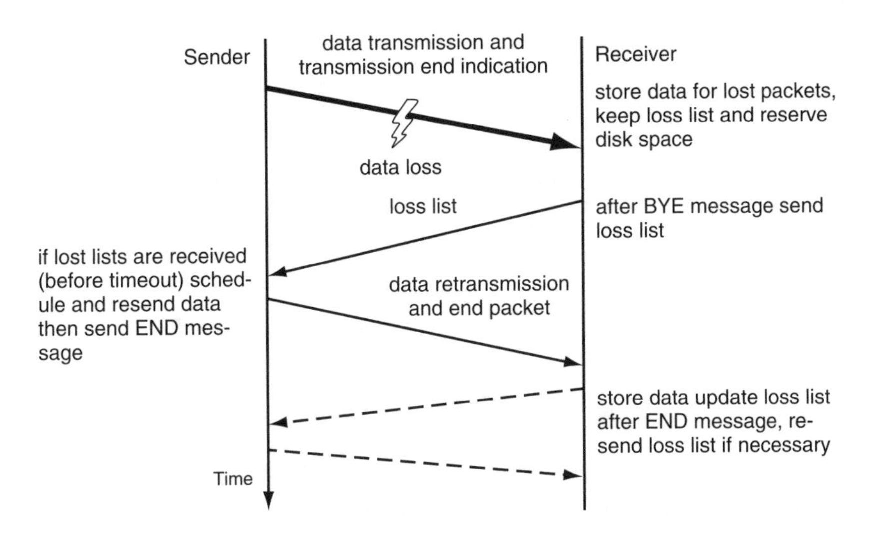

Figure A.1 LC-RTP communication.

client, while with the multicast method data can be streamed in parallel to several caches and clients (see Figure A.2). Additional measures that had to be taken to allow multicast in combination with LC-RTP are explicitly outlined in the following sections of this appendix.

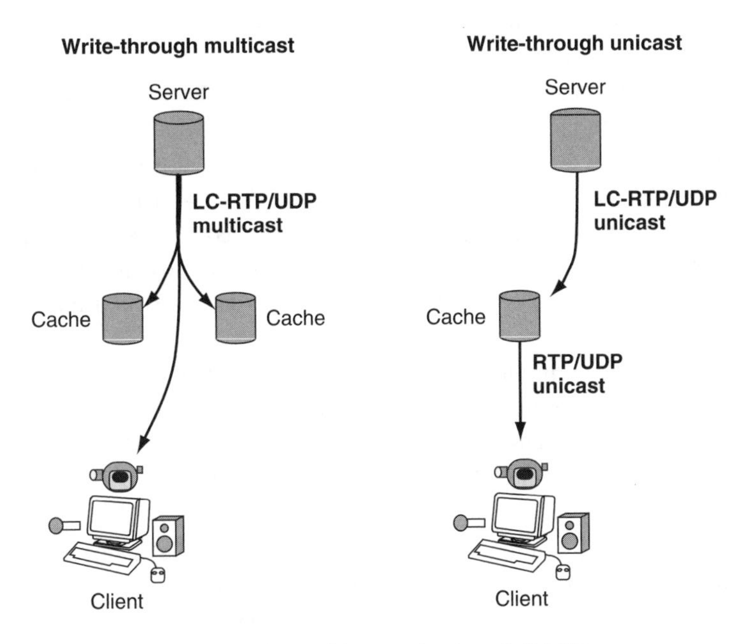

Figure A.2 Multicast and unicast LC-RTP.

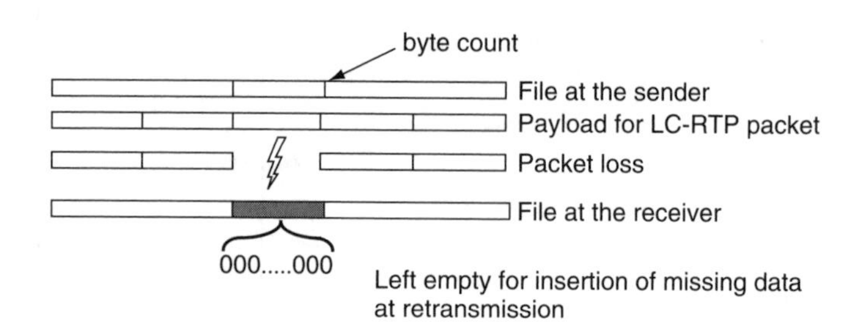

Figure A.3 LC-RTP byte count supports retransmission.

A.3.2 Actions during the Content Transmission

The sender streams the content that is requested by a client as a multicast stream to all receivers of a multicast group including the clients. In order to give the caches the opportunity to reserve exactly the required disk space in case of data loss, it is necessary to send information beyond the regular information of a RTP packet. In the case of LC-RTP this information consists of a byte count which is included in each RTP packet. This mechanism facilitates the synchronization between byte count and the data which is represented by it. If the byte count were sent in an extra packet, e.g. via RTCP, the sequence of the byte count and data packet can be interchanged, or the byte count packet can get lost.

The receiver stores the data and detects a loss by checking the byte count with the last memorized byte count. If a packet loss is detected, the difference between the two byte counts and the length of the actual packet is computed and this computed size can be reserved on the disk for a later insertion of the retransmitted data (see Figure A.3). The received payload of the packet is then stored after this reserved gap. Furthermore, the loss must be written to a loss list. If no loss is detected, the received data is stored on the disk immediately.

Each cache server implementation has to transform the byte count value into its own file indexing information. As a consequence it is possible to have different file layouts on the server and cache. For example one cache implementation stores the file as raw data and another stores some header information with it.

As a consequence of including the byte count in the data packet, and the requirement of serving regular RTP clients, only an RFC-conforming protocol extension was an option; including the byte count in the RTP payload of the packet would cause problems for standard receivers (see section A.4).

At the end of the transmission, an end packet is sent including the last byte count, in order to inform the receivers of the normal end of the transmission including information to check whether data preceding the end packet was lost.

Reserving the computed space in the file in the case of a loss detection has several advantages. The approach of reserving the correct amount of space on the hard disk is very simple and efficient, because it preserves the sequential nature of the stored data. This property is essential for an efficient use of a hard disk, as seeking on a disk significantly diminishes its throughput. Furthermore, this allows LC-RTP to be compatible with multimedia file systems [167, 168] which are penalized by inserting or do not support it at all.

A.3.3 Actions after the Content Transmission

After sending the end packet the sender starts a timer. This timer should be a multiple of the worst-case RTT (round trip time) between the sender and the known receivers. This RTT can be computed with the periodic RTCP packets that are sent for calculations of the network quality. During this timer period at least one loss list has to be received from a cache that has detected packet losses, or the session ends.

With the reception of the end packet the cache finishes the normal procedure of the transmission of the content and starts the procedure for initiating retransmissions. To avoid a possible overload of the sender, loss lists are sent from receivers after a random amount of time in the case of multicast LC-RTP. The loss list includes all ranges of the detected data losses. If ranges are direct neighbours, they are combined into one range, in order to keep the size of the list small.

If a loss list arrives at the server, the requested data ranges are stored in a schedule list. This list includes a counter for each range to indicate the number of requesting clients. This allows the use of a strategy for building a retransmission schedule (for example, most frequently lost packets first).

Resent packets are of the same size as the packets that were sent during the first transmission to simplify storing at the receiver. The resent data range is deleted from this list. The client saves each requested, retransmitted packet at the position that is indicated by the byte count. Concurrently, the loss list is updated. If the byte count is not included in the loss list the packet is discarded.

When the last entry of the list is processed and deleted, the sender resends the end packet in order to inform the receivers that this retransmission cycle is over. This procedure is repeated until an application-specific retransmission counter has reached its threshold value or until no more loss lists are sent.

To avoid the blocking of a receiver a timer is necessary that terminates the session if no end packet or other resent packets are received after a considerable period.

A.4 USE AND INTEGRATION OF PROTOCOLS

The design of LC-RTP was made within the constraints of an RFC-conforming RTP implementation. This section describes how the standard RTP protocol is extended to meet the goal described above.

A.4.1 LC-RTP as an RTP Extension

The main problem in mapping LC-RTP into RTP is the byte count, as it has to be included into the header of RTP (see section A.3). This is necessary in order to keep the content of LC-RTP packages compatible with RTP-related packaging RFCs and, therefore, to make it possible for standard RTP clients to receive LC-RTP streams. One way of inserting the byte count into the RTP header and not into the payload is to use the extension header of RTP (Figure A.4). By setting the extension bit, a variable-length header extension to the RTP header is appended. LC-RTP defines two kinds of header extensions. They are defined to easily distinguish whether a packet is sent as part of the regular stream or during a retransmission phase. The only difference between them is the value in the identifier field. Each extension

```
0                   1                   2                   3
0 1 2 3 4 5 6 7 8 9 0 1 2 3 4 5 6 7 8 9 0 1 2 3 4 5 6 7 8 9 0 1
+-+-+-+-+-+-+-+-+-+-+-+-+-+-+-+-+-+-+-+-+-+-+-+-+-+-+-+-+-+-+-+-
+
|       defined by profile          |            length        |
+-+-+-+-+-+-+-+-+-+-+-+-+-+-+-+-+-+-+-+-+-+-+-+-+-+-+-+-+-+-+-+-
+
|                         byte count(64 bit)                    |
+-+-+-+-+-+-+-+-+-+-+-+-+-+-+-+-+-+-+-+-+-+-+-+-+-+-+-+-+-+-+-+-
```

Figure A.4 RTP header extension.

header consists of the two RTP-dependent extension fields plus an additional byte count field. For a current video streaming application this field should be 64 bits long, as a cyclic byte count must be prevented.

During the usual transmission, the RTP transmission is performed regularly, except for the byte count which is included in the RTP extension header. At the end of the transmission an end packet is sent. An appropriate way to do this is by sending an RTCP packet. This packet should not be the normal RTCP BYE packet, as this is used for other meanings. So an application-dependent extension RTCP packet must be created, as shown in Figure A.5.

LC-RTP defines two application-defined RTCP packets. The first one is the end packet and the second one is the loss list packet. The only additional data transmitted in the end packet is the last byte count of the session. The name of the packet itself is enough information for the receiver to interpret this as the end of the normal transmission. The list appended into the loss list packet should be appended as a list of byte count ranges.

```
 0                   1                   2                   3
 0 1 2 3 4 5 6 7 8 9 0 1 2 3 4 5 6 7 8 9 0 1 2 3 4 5 6 7 8 9 0 1
+-+-+-+-+-+-+-+-+-+-+-+-+-+-+-+-+-+-+-+-+-+-+-+-+-+-+-+-+-+-+-+-+
|V = 2|P| subtype |    PT = APP = 204  |            length        |
+-+-+-+-+-+-+-+-+-+-+-+-+-+-+-+-+-+-+-+-+-+-+-+-+-+-+-+-+-+-+-+-+
|                           SSRC/CSRC                             |
+-+-+-+-+-+-+-+-+-+-+-+-+-+-+-+-+-+-+-+-+-+-+-+-+-+-+-+-+-+-+-+-+
|                  name (ASCII)   (set to LRTP)                   |
+-+-+-+-+-+-+-+-+-+-+-+-+-+-+-+-+-+-+-+-+-+-+-+-+-+-+-+-+-+-+-+-+
|                  application-dependent data ...                 |
+-+-+-+-+-+-+-+-+-+-+-+-+-+-+-+-+-+-+-+-+-+-+-+-+-+-+-+-+-+-+-+-+
```

Figure A.5 Application-defined RTCP packet.

The extension to RTP is minimal and should be ignored by other applications. This is very important, because it ensures that a cache update can be made in parallel with a client request. (Experiments with LC-RTP in combination with *vic* and *vat* resulted in the rejection of all RTP packets with an extension header. A closer look at the source code of both revealed that the RTP implementation is not standard-compliant.)

For the intended application class, the header extension introduced by LC-RTP is sufficiently cheap with an overhead of 8 to 12 bytes per packet. Furthermore, this type of extension is defined in the original RTP RFC [62] and should – theoretically – be implemented by all RTP implementations.

A.5 TESTS

RTP and LC-RTP were implemented in the KOMSSYS streaming platform. This implementation was used for an Internet-based experiment which is described in detail in the following.

A.5.1 Test Scenario

The goal of this experiment is to show that LC-RTP performs as well and reliably as other data distribution protocols (e.g. FTP) and can be used for the reliable distribution of A/V content via both unicast and multicast.

Two different video objects (6 MB and 20 MB of MPEG-I Movie) were transmitted from locations in Germany, the USA, and Canada to a receiver located at our institute (Darmstadt, Germany). The results from transmissions between the USA (National Institute of Standards and Technology) and Canada (University of Ottawa) (both acting as senders) and a receiver located in Darmstadt are presented in Table A.2. Single experiments were repeated five times for each file from both locations, each time with a different transmission bandwidth.

The decision to perform long-distance experiments was made because of the higher likeliness of packet losses. During preliminary experiments on the campus network and within Germany, no losses, or only very few losses, occurred.

For each LC-RTP session information about the retransmission was logged at the receiver and the original file and the transmitted file were compared

Table A.2 Test results (bandwidth, duration)

		Max. BW (bit/s)		Duration (s)	
BW (kbit/s)	File size (Mbyte)	NIST	Ottawa	NIST	Ottawa
1 000	6	1 047 552	1 022 800	41	42
	20	1 024 048	1 024 000	160	160
2 000	6	2 147 480	2 045 216	20	21
	20	2 048 104	2 048 512	10	11
4 000	6	4 294 968	3 904 512	10	11
	20	2 561 080	4 096 000	105	40
8 000	6	8 593 216	1 169 880	5	37
	20	8 192 008	1 058 392	20	151
12 000	6	8 589 936	1 213 296	5	36
	20	5 461 336	487 968	30	337

to assure that the transmission completed successful. The comparison for all tests was positive, that is all transmissions were finally made without any errors.

A.5.2 Test Results

The results obtained from the logging that was performed during the LC-RTP sessions shows the occurrence of retransmissions in almost all of the tests. The logging information also confirmed that the number of retransmissions increases with the size of the bandwidth it has tried to send the files with. If the bandwidth is set much higher than the actual bandwidth of the link between sender and receiver, multiple retransmissions for one packet are more likely. However, in these cases too the files were transmitted without any errors.

Table A.3 Test results FTP

	Max. BW (Bit/s)		Duration (s)	
File size (Mbyte)	NIST	Ottawa	NIST	Ottawa
6	576 000	328 000	71	126
20	568 000	204 000	273	512

During the tests it also became clear that the quality of the link between the USA and Darmstadt is of a higher quality than the one between Canada and Darmstadt. We also transmitted both files via FTP from both locations to Darmstadt to obtain some information about the performance of a traditional file transfer protocol. The comparison of the transmission times shows that, with LC-RTP, data can be transmitted faster than with plain FTP. This is caused by the nature of the UDP-based transmission which does not, in comparison to TCP, back off in the case of congestion in the network. Thus, the performance of competing TCP traffic is affected by LC-RTP transmissions in the case of congestion. In section 8.4.1, a TCP-friendly approach for LC-RTP is presented.

A.6 SUMMARY

Caching and prefetching of A/V content are powerful methods of increasing overall performance in the Internet. LC-RTP is a simple and efficient reliable multicast protocol compatible with the original RTP, which is stated by the experiment presented in this appendix. It needs to be implemented only in servers and caches. These servers have to be adapted to LC-RTP and they need mainly a list implementation, so the adaptation is a very simple procedure. Clients are not affected at all by LC-RTP.

All resources are used carefully and the extension permits an implementation to use a simple method to keep the sequential nature of the stored data without buffering. This method considers hard disk performance and possible network structures without wasting resources (such as main memory and CPU power). Its intention is to allow a maximum number of concurrent streams handled by the caches. As no additional packets are sent during the regular session and the packet sizes are hardly bigger than those of an standard RTP sender, all access control mechanisms and network quality computations can remain unmodified. The only difference to a normal transmission is the fact that, after the session, a retransmission of the lost packets to between server and cache with LC-RTP extensions is performed. A conforming, standard RTP receiver would recognize this as a normal session termination and, thus, would not be affected.

Multicast ensures a minimum load increase on the network, because the packets are sent only to members of the multicast group, during a transmission to a regular customer.

Appendix B: Preliminary Subjective Assessment

The main goal of the subjective assessment from Chapter 4 is to get an answer to the question: How do variations in a layer-encoded video influence its perceived quality? Since the answer to this question should have an influence on the work on retransmission scheduling, the main idea is to compare impaired sequences directly with each other. In general, two of the presented test methods from section 4.3.3, double stimulus impairment scale (DSIS) and stimulus comparison (SC), are applicable in this case. With DSIS information about the influence of an impairment in comparison with a reference sequence can be obtained. Thus, the comparison of two impaired sequences can only be performed indirectly. For example, to compare two impaired sequences (sequence 1 and sequence 2) two DSIS tests have to be performed. In the first test, sequence 1 and the reference sequence are compared while in the second test a comparison between sequence 2 and the reference sequence is made. In contrast to DSIS, the SC method needs only one single test for the comparison of sequences 1 and 2.

Since it was not clear which of the two tests should be used for the subjective assessment, the decision was made to perform a preliminary assessment in order to find out which method is suited best. An additional goal of this test was to investigate whether the test candidates would recognize the quality changes that occur in the test sequences. Negative results on this investigation would have implied a modification of the test sequences.

B.1 EXECUTION OF THE PRELIMINARY ASSESSMENT

This assessment was executed manually, since the test application described in section 4.3.4 should make use of the results gained in this investigation and naturally could not be available at this point in time. (The implementation of

Scalable Video on Demand: Adaptive Internet-based Distribution M. Zink

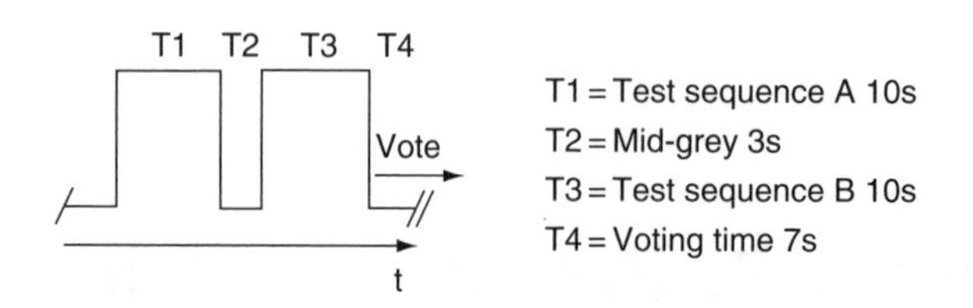

Figure B.1 Single test procedure.

the application should not start before the final decision on the test method has been made in order to avoid implementing two methods.)

The procedure of the test was as follows. First of all the test procedure was explained to the test candidate. The questions and possible answers for each single test were given to the candidate. The candidate was given sufficient time to read and understand all questions and answers. Afterwards the actual test sequence was presented to the client with the time constraints as shown in Figure B.1. Figure B.2 shows the sequences used for the assessment.

After the presentation of the two sequences the candidate had to answer the question in the next 7 seconds by marking his assessment. Six single tests were executed during the whole assessment in which 14 test candidates took part. The candidates should compare M&C 1 with M&C 2 and T-Tennis 1 with T-Tennis 2 using both test methods. With the DSIS test method four single tests are necessary, while the SC method requires only two tests. A final comparison between the two test methods seems to be valid, since equal content and quality variations were used for each test method. The way the candidates assessed the quality of the sequences is shown in the following two sections.

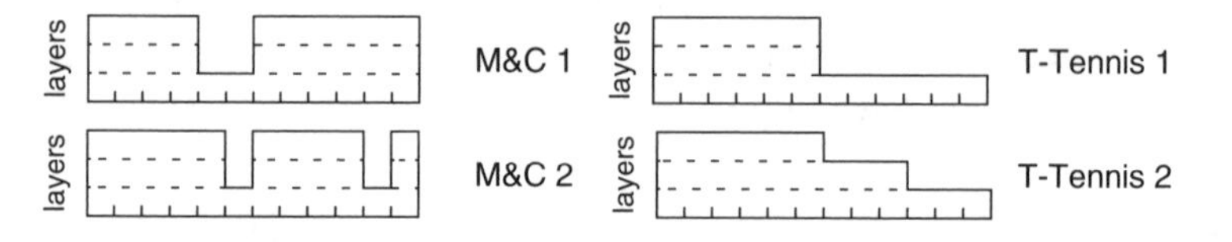

Figure B.2 Shapes for the preliminary assessment.

B.1.1 DSIS Method

The same video sequence will be shown to you twice. The quality of the two sequences ***may*** *differ. Please, answer the following question (within 7 seconds)* ***immediately after the second sequence****.*

Did you perceive the second sequence in a worse quality than the first and if so was it annoying?
(You have to answer this question within 7 seconds.)

	imperceptible
	perceptible, but not annoying
	slightly annoying
	annoying
	very annoying

B.1.2 SC Method

*The same video sequence will be shown to you twice. The quality of the two sequences **may** differ. Please, answer the following question (within 7 seconds) **immediately after the second sequence**.*

Mark on the shown scale (shown below), within 7 seconds after you have seen the second sequence, how you perceived the 2 sequences relative to each other.

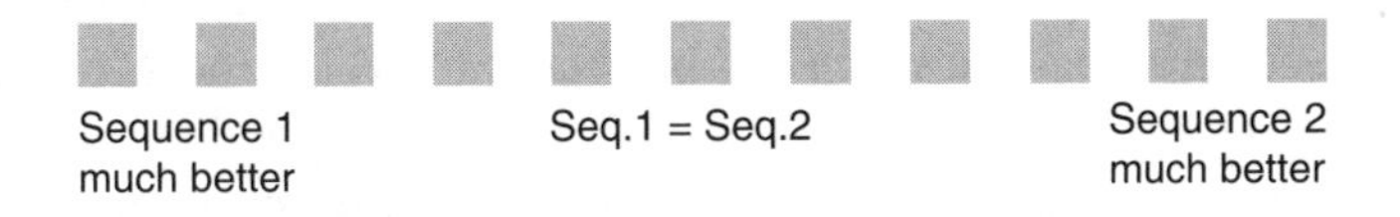

B.2 SELECTION OF THE TEST METHOD

The outcome of the statistical analysis (performed as described in section 4.5) of this preliminary assessment is shown in Figure B.3.

The first positive results of the assessment is the significance of the statistical analysis. The tendencies that can be derived for the single tests met the expectations. First, the tests with the M&C sequence as content for the experiment are discussed in more detail. The results of the statistical analysis (two DSIS and one SC) are shown in Figure B.3. M&C 1 shows the result of the DSIS method where the original sequence and sequence M&C 1 (see Figure B.2) had to be compared with each other, while M&C 2 shows the result of the comparison between the original and the M&C 2 sequence. M&C 1/2 presents the result of the SC test at which sequences M&C 1 and M&C 2 are compared directly with each other. The negative result of this test indicates that the test candidates assessed the quality of the first sequence

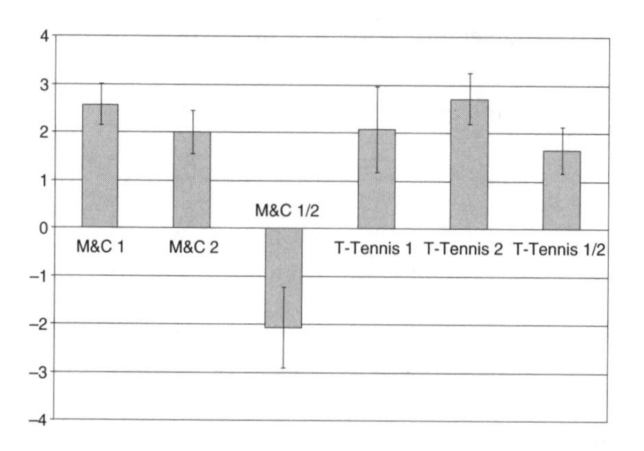

Figure B.3 Average and 95% confidence interval for the different tests of the experiment.

(M&C 1) better than the second (M&C 2). In addition to the average result of the assessment the 95% confidence interval of the T-test is also shown. For M&C 1/2 the confidence interval is quite large but does not cross the neutral axis. Therefore, it is of statistical significance. The same is true for the results of M&C 1 and M&C 2. The test candidates noticed an impairment in the modified test sequences, yet the results for both tests are quite similar. To be able to compare the results of both tests (M&C 1 and M&C 2) the difference between both results must be calculated. The confidence intervals of both tests overlap and, therefore, are not as significant as the result of the single SC test. If the average values only of M&C 1 and M&C 2 are regarded, a tendency similar to the result of the SC test (M&C 1/2) can be recognized.

Similar results were obtained for the second pair of sequences compared with each other. While the result of the SC test method (T-Tennis 1/2) is statistically significant, the conclusions that can be drawn from the two DSIS tests (T-Tennis 1 and T-Tennis 2) do not allow such an interpretation. Regarding only the average of the assessments reveals an equal tendency for both test methods: T-Tennis 2 is assessed as being of higher quality than T-Tennis 1.

The initial goal of this preliminary assessment was to make a decision on which of the two test methods (DSIS or SC) should be used in further assessments. First the advantages of both tests are given in the following and then the decision for the SC test is justified.

- In comparison to the DSIS method the results of a statistical analysis based on the SC method can be statistically significant.

- The SC method is well tailored for the goal of this investigation: two sequences, each having a different, impaired quality should be compared with each other directly.
- With the DSIS method the mean value of one test can be located within the boundaries of the confidence interval of the other (peer) test and vice versa. This effect can lead to an overall increase of the confidence interval and, thus, the possibility of obtaining statistically significant results is reduced.
- Compared to DSIS the amount of context effect [169] is reduced. Thus, results of the SC method are more stable.
- The overall duration of the assessment is shorter with the SC method, since only half the number of tests are required compared to the DSIS method.

The DSIS method has the following advantage:

- ITU-R BT.500-10 [136] states that the DSIS method is well suited for a comparison of the original and a sequence in which impairment is small. For example, quality degradation is due to bit-errors occurring during the transmission of the video rather than to quality degradations that are introduced due to an adaptive streaming.

In general, the goals of the two test methods are slightly different. While the DSIS method is used to assess more fine-grained impairment, SC aims at more general statements about the quality of a video sequence. This is also reflected by the two different scales (see sections B.1.1 and B.1.2) to assess the single tests. Nevertheless, this difference should not influence the decision in favour of one of the other test, since both methods can produce meaningful results.

The decision to use the SC method was taken because of the number of advantages it offers in comparison to the DSIS method. The most important fact is the possibility of obtaining statistically significant results more easily than would be the case with DSIS. The only benefit of the DSIS, assessing the quality of an impaired sequence in comparison to a full quality sequence, is not necessarily needed for the assessment that should be performed in this book. And the fact that the duration of a complete assessment is doubled by the DSIS method should not be neglected.

B.2.1 Content

In discussions that followed the assessment, test candidates stated the influence of the content on the way they performed their assessment. According to many of the candidates, a sequence is assessed differently when the content of this sequence is watched for the first time compared to later assessments when the content is already known. This effect could be caused by the fact that new content is distracting the test candidates. That is, test candidates are concentrating on the content instead of the impairment of the sequences. To avoid this phenomenon, following reference [169], initial sequences were included in further assessments (as in section 4.4). Those test sequence make the user aware of the content but are not used for further statistical analysis.

Appendix C: A Toolkit for Dynamically Reconfigurable Multimedia Distribution Systems

C.1 MOTIVATION FOR A VIDEO DISTRIBUTION TESTBED

In recent years a substantial amount of work has been performed on investigations of wide-area audio and video (A/V) distribution in the Internet. Most of this work has been theoretical and simulative work on new mechanisms that reduce the consumption of network and server resources by streaming data in A/V distribution systems (see Chapter 3). During work on transport mechanisms for such systems we realized that a testbed for implementing the mechanisms that are used on the data path is necessary for measurements and analysis to be performed. Therefore, the decision was made to build a toolkit for the testing of A/V streaming and distribution mechanisms (KOMSSYS, [162]). With the knowledge that the development of this infrastructure would be performed as a research project mainly at universities and, therefore, would be supported by contributions from student projects (master's degree theses), the system should be easy to extend and have reusable components and well-defined interfaces. Therefore, a toolkit was created that allows implementors to build prototypes for multimedia distribution systems. Such a distribution system prototype comprises simple applications that fulfil the basic functionality of video servers, video caches and clients. These applications can be built from the toolkit or they can be existing applications such as the RealPlayer [51] or the Quicktime Streaming Server [170]. The components of the toolkit provide abstractions for data

Scalable Video on Demand: Adaptive Internet-based Distribution M. Zink

and protocol handling functions that are required on the data path of such distribution systems. To allow interoperability with existing applications, the networking components of the toolkit were built around a standard-based architecture that supports the existing standards RTP/RTCP [62], RTSP [66] and SDP [65].

Existing approaches for configurable distributed multimedia systems [171–176] implement mostly data path components that are connected into graphs which remain unchanged while data is flowing, or they consider middleware frameworks that allow the specification of an end-to-end behaviour for complex multimedia systems. In the latter kind of systems, functionality is described at the level of cooperating distributed components [177, 178]. Achieving network transparency is neither a goal nor a possibility of the toolkit. Firstly, third-party applications are included in the distribution systems; secondly, standardized and extended protocols for the data path should be investigated. These standards require support for reconfigurations of the data path. Reconfigurations are necessary because the data path can be influenced by control information that is implicitly contained in protocols and payload. In RTP, for example, packets may arrive unpredictably from new sources, or MPEG transport multiplexes may contain other streams than those expected.

In the following, the toolkit design, its performance aspects, and the experience gained during its use, by the original developers and others who have implemented several distribution system mechanisms, are presented. Experiences include the example of the gleaning distribution mechanism [4], which was implement by a graduate student [179] by adding just two new components for the data path. Performance measurements with both a monolithic and modular architecture for the toolkit showed that the performance penalty for the latter approach is only around 8%.

C.2 TERMINOLOGY

Several terminologies have been used in the past to describe multimedia distribution subsystems. The *stream handler* terminology has been used for many years and is also adopted for the toolkit presented in this appendix. The example for a data path that is set up in a simple streaming server is used to explain several terms and their functionality, as shown in Figure C.1. In this specific example three components are used that read video data (in this case MPEG-1 System) from disk, packetize it and add RTP header

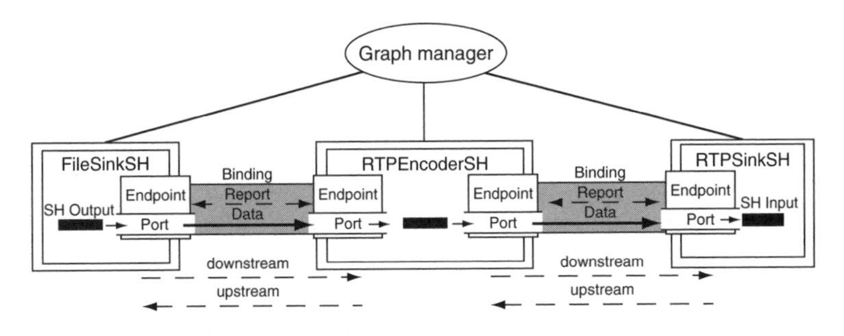

Figure C.1 Terms.

information, and send it out on the network. Based on this example, several terms are explained in more detail in the following:

Stream handlers (SH) are components that can be bound together dynamically by a controlling entity into a graph. A SH must either produce or consume data units, or data units can enter and leave the SH. Data units that enter the SH are *SH input*; data units that leave are *SH output*. A *port* of an SH is an interface for sending or receiving data units. An *endpoint* encapsulates ports of a SH. Endpoints provide the information that allows one to decide whether ports of two SHs can be bound to each other or not. Only if one SH's output port is bound to another SH's input port can data units be exchanged between the SHs. These *bindings* are logical entities in our case, implemented as function calls. The terms *upstream* and *downstream* are meaningful because the resulting graphs are directed. Upstream qualifies SHs that are bound to an SH's input port; downstream qualifies SHs bound to an output port.

The graph shown in Figure C.1 consists of three SHs, *FileSinkSH*, *RTPEncoderSH* and *RTPSinkSH*. In this example, data is only sent downstream from the *FileSinkSH* through the *RTPEncoderSH* to the *RTPSinkSH*. Control data, e.g., RTCP information, can be sent upstream and downstream between the SHs. Data units are, for example, video data that leave *FileSinkSH* as *SH output* and enter *RTPEncoderSH* as *SH input*.

A *stream graph* (or simply *graph*) is a set of connected SHs and the bindings between them. In graph terminology, we consider the SHs nodes and the bindings edges. Here, the movement of data units in stream graphs is directed and non-cyclic. *Sub-graph* describes a subset of SHs that are bound directly to each other but have unbound ports. A *trunk* is a sub-graph of a stream that has either only open input or open output ports.

The controlling entity mentioned in the SH definition is called the *graph manager (GM)*. On behalf of an application, the GM creates a graph of SHs to form a media-processing subsystem. The GM is responsible for the setup and destruction of the SHs, determines the interaction between the individual SHs and is the interface to the application. Since the focus of this work is on the data path of distribution systems, these GMs are application-specific. They use predefined sub-sequences of SHs that are required for a specific task, such as data forwarding, writing to and playout from disk, buffering, or sequencing. The applications and the subsystem communicate only through this GM.

In the toolkit, a *stream* consists of data that is logically one entity and that is processed by a single stream graph. The same stream graph may handle several streams in parallel.[†]

C.3 DESIGN

KOMSSYS, the streaming platform that makes use of the toolkit, offers the possibility of making practical experiences with various distribution system mechanisms. This requires that mechanisms that are implemented using the toolkit can achieve comparable performance to monolithic, specialized implementations. The toolkit must allow the creation of mechanisms in such a way that run-time performance is equivalent to a dedicated implementation. This goal requires also the ability to reconfigure the graphs that are built from the components of the toolkit, because distribution mechanisms require changes to the data flow that can be handled internally by monolithic implementations, but require reconfiguration in a graph of interconnected components.

Since the toolkit should allow other researchers to build prototypes for multimedia distribution systems, it must be easy to extend and have reusable components and well-defined interfaces. In section C.3.3, we describe how the toolkit supports the implementor in building new prototypes.

C.3.1 Equivalent Design

The prototype applications that build on the basis of the toolkit should allow investigations on the performance of the data path. Therefore, the implementation should have a low influence on this performance, i.e., the toolkit should not force an implementor of a mechanism to build a solution that consumes more system resources than a monolithic implementation of the same mechanism. Such basic limitations would arise from a toolkit-defined

[†] In the remainder of this chapter 'flow' is used as a synonym for 'stream'.

threading model and memory model that are mandatory for all components. These influence fundamental issues in approaches such as the amount of queueing of data and control information, and the number of mandatory context switches. Besides the support for passing arbitrary structures of data between components after an initial negotiation of capabilities, the threading model and the handling of control information are largely determined by these performance considerations.

C.3.1.1 Threading Model

The threading model of an SH system is concerned with the question, which entity is processing what amount of data at what time. Advanced, open middleware approaches that implement functions by concatenating functional modules into arbitrary graphs of independent components are able to attach scheduling mechanisms to arbitrary sub-graphs [180]. While this approach is highly flexible, it requires either an operating system abstraction layer to allow arbitrary grouping, or information about the potential grouping capabilities of modules. For example, it is problematic to support a module that listens to BSD sockets with a module that waits for the release of a POSIX semaphore in the same thread.

The toolkit consist of both, very fine-granular SHs with minimal functionality, such as logically copying a data unit, and course-granular ones, such as an SH that controls a zero-copy kernel implementation [181] which handles everything from data retrieval from disk to sending onto the network. The first example makes it undesirable to have one thread per SH; the second requires that threading can be controlled by SHs. To fulfil both requirements the optional creation of threads inside each SH and also optional sharing of threads among SHs with assistance of the GM is supported.

To organize the creation of graphs from SHs that have various demands for concurrency, they can be referred to as active-capable, passive-capable, and through-capable SHs. The terms 'active', 'passive' and 'through' are defined as follows:

- *Active*: Active SHs determine their own timing, usually by waiting for events of some kind, such as timeouts or packet arrivals. They can push data downstream actively (by calling a push function of the downstream SH) or they can pull data from an upstream SH, or both. Two active SHs cannot be connected directly because each one tries to control synchronicity. A passive SH must be inserted between them. The *RTPEncoderSH* in Figure C.1 acts as an active SH: it pulls data from the *FileSinkSH* and, based on its own timer, pushes it to the *RTPSinkSH*.

- *Passive*: A passive SH does not determine timing. If it acts as a sink, an upstream SH may push data to it; if it acts as a source, a downstream SH may pull data from it. If it implements both source and sink, it must also provide buffering capacities that suit the needs of the graphs that it is likely to be included in. Either in-band reporting or GM notifications can be used to warn of over- and under-runs of the buffer. Passive SHs cannot be connected directly because no data would be exchanged between them. *FileSinkSH* and *RTPSinkSH* act as passive SHs in Figure C.1.
- *Through*: Through SHs are meant for tasks such as on-the-fly transcoding, packet duplication, or filtering. They do not generate timing and should not introduce buffers beyond those necessary for their operation. They must always implement a source as well as a sink. An arbitrary number of them can be concatenated. An active SH that is located upstream will push data through this kind of SH, potentially through several more through SHs until a passive SH is encountered. The pull operation is used in the same way by an active SH located downstream. Whether a through SH can be used in push or pull mode is determined by endpoint settings. The matching function of the endpoints can restrict it to one direction. In section C.4.4 an example for a through SH is given.

A SH may be capable of one or more of these operation modes. During the configuration or reconfiguration of a stream graph, the matching procedure must determine whether the desired sequence of SHs can be connected without violating these threading conditions. The mode of an SH that supports several or all of the three modes can only be determined when it is initially added to a graph; but it remains fixed during reconfiguration.

Independently of these modes, each SH is allowed to create as many threads internally as are necessary for its processing. Only if these threads perform data retrieval from upstream SHs or data delivery to downstream SHs do the SHs identify themselves as active SHs to the matching process that is initiated by the GM during configuration and reconfiguration. Otherwise, the GM remains oblivious of internal threads. This approach limits control over thread use in the system, but it simplifies the integration of third-party modules, even closed-source modules (see section C.4.5). To allow better control and a reduction of the number of concurrent threads when necessary, SHs can on the other hand offer an interface to the GM to provide them with a thread object, which can then be shared among several SHs at the discretion of the GM.[†]

[†] Optionally, a single-threaded approach for simplified debugging only is also supported, which is, however, not generically applicable to all SHs.

A large number of other options were considered but discarded for the following reasons. Using one thread per graph has the disadvantages of a single thread for the entire system but the additional problem of losing control over the complete system. An approach that assigns one thread to each data packet wastes threads when processing is delayed, and it fails when packets are copied, split or merged. Running each SH in a separate thread can encapsulate third-party software well, but a simple packet forwarding in a graph requires an unnecessary number of context switches. Special thread SHs provide good speed control to the GM but require that all other SHs are passive and adaptive to speed changes.

C.3.1.2 Feedback and Parameterization

SHs have been devised to ease the integration of streaming media into applications. The original assumption is that SHs are connected and configured at application startup, resource needs are negotiated for all relevant components, and the streaming is than handled by the SH graph transparently. State-of-the-art multimedia systems support resource-adaptive media processing, for example in the control-theoretical approach, where a set of SHs cooperates logically to form a control loop [182]. This requires the distribution of feedback information to the SHs that must adapt according to this information.

The toolkit allows two different types of feedback distribution, one that allows the exchange of feedback between SHs directly, and one that allows it between an SH and the GM. The direct exchange between SHs (in-band) saves resources and context switches when SHs generate a large amount of feedback information and the number of existing SHs in a stream graph is high. Nevertheless, there are cases when the application must be notified and, therefore, feedback information can also be sent to the GM (out-of-band). Acceptance of this feedback will usually require context changes.

In the in-band approach, each SH implements a return channel that takes control messages from a report interface and tries to interpret and process them. If they cannot be interpreted, they are forwarded to all SHs that are connected to the other side of the SH, i.e., to all upstream SHs if the report arrived from the downstream side and vice versa. If they can be interpreted, the SH may be able to complete the processing, it may decide to notify the GM, or it may forward a modified report to the upstream SHs. Since the GM has created the stream graph, it can change the parameters of the relevant SHs directly, bypassing potential intermediate SHs. The in-band approach is most appropriate for sets of SHs that are usually all present in a stream graph and that cooperate in a predefined manner, whether they are separated by other SHs or connected directly. The out-of-band approach is more useful for the handling of exceptional situations or atypical configurations, or for the delivery of feedback over machine boundaries, for example, signalling information from RTSP that requires a modification of the behaviour of one or several SHs.

An alternative approach, the handling of control information in an additional graph of SHs, was not pursued because of the complexity it would add. This approach would double the number of SHs that need to be managed and it would introduce cycles in the SH graph.

C.3.1.3 SH Performance

The SH architecture of the toolkit requires that data is forwarded between arbitrary SH endpoints. This implies a dynamic identification of data handling functions which is realized through the typical C++ approach of virtual functions. To evaluate the efficiency of this approach, a pipe of three dynamically plugged SHs is compared with an SH that hard-codes the same functionality in a single module. The experiments are performed on an unloaded system, but in multi-user mode. The consumed time is measured in CPU cycles using the Pentium-specific mnemonic RDTSC [183]; the compiler is a GNU g++ 2.95.3 (with -O2) on Linux 2.4.14. The chosen encoding format is a dummy format that does not require disk access when data is 'read from file' to prevent disk reads falsifying the measurement.

Figure C.2 shows the performance increase that can be achieved by merging a pipe into a single, specialized SH. In the pipe comprising three SHs, the RTPEncoderSH is an active SH, while the other two SHs are passive. When the streaming graph is active, the encoding function of the RTPEncoderSH calls a referenced, codec-specific object at times that are determined by the RTP packaging format for the specific encoding. This codec checks its prefetch buffers (*code a*) and if it needs additional data, pulls chunks of data from the FileSourceSH via the sink endpoint of the RTPEncoderSH and the source endpoint of the FileSourceSH (*pull*). Then, it processes the data chunk (*code b*) and returns it to the RTPEncoderSH. It is then sent to the RTPSinkSH via the source endpoint of the RTPEncoderSH and the sink endpoint of the RTPSinkSH (*push*).

The chart on the left side (Figure C.2) shows single function calls, while two virtual function calls and an indirection through source and sink endpoints are performed on the right side. The time for each call is denoted *enter code*, *enter pull*, and *enter push*, and it contributes an essential portion

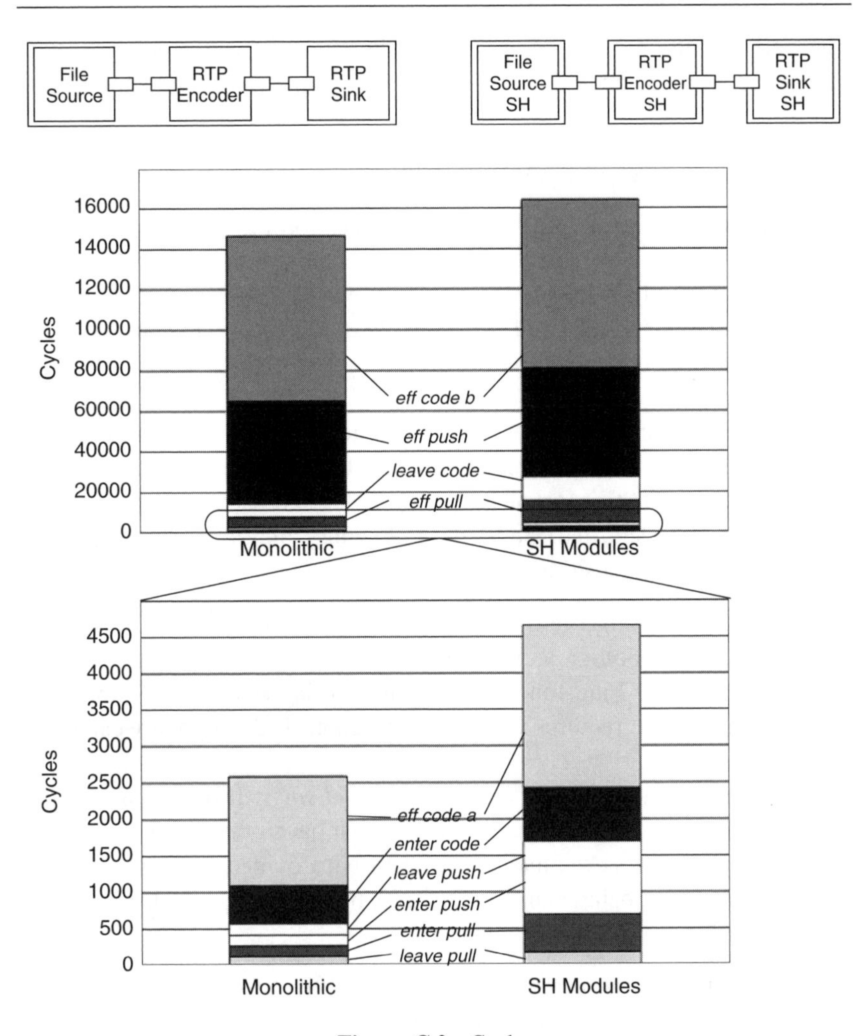

Figure C.2 Cycles.

of the performance penalty on dynamic pipes of SHs. While this penalty was expected, it was unexpected that there is also a considerable penalty in leaving virtual function calls, at least partly due to conservative smart pointer handling.

The hard-coded SH saves time in three ways: by replacing virtual function calls with direct function calls, by increasing data locality due to collecting all member variables in a single object, and by implementing all functions within a single object instead of separate objects for SHs, SH endpoints and codecs, and by additional smart pointer construction and destruction.

The performance penalty of approximately 8% in a nearly empty pipe is acceptable when compared to the inflexibility of the integrated approach.

C.3.2 Streaming Graph Reconfiguration

In architectures that use coarse-granular SHs for the creation of stream graphs, such as the one used for the toolkit, reconfiguration of active stream graphs is frequently a matter of reconfiguring the behaviour of individual SHs rather than the graph itself. An example of the options that are available through this design is given by [182] in a system that is adaptive through feedback control loops.

Many applications do not need any dynamic reconfiguration of data paths and, therefore, do not require reconfigurable stream graphs. In the following, cases in which modification of individual SHs' behaviour will solve many requirements are identified. Under the following conditions, reconfiguration of the graph is an interesting option:

- *Long life-time*: Long life-time of a stream graph makes the necessity of changes in the topology more likely.
- *Fine granularity*: Fine functional granularity increases the flexibility of graph creation but requires very complex graphs if the topology cannot be changed.
- *Unpredictable behaviour of data sources and sinks*: If data sources, such as an RTP sender, can produce data units that have other processing needs than earlier units, new functions have to be provided. Data sources can also change their behaviour, for example, by modifying the display speed or requesting a different data format.

The toolkit supports reconfiguration of the active stream graph, meaning that data continues to flow through the graph while sub-graphs are added to or removed from the graph. These operations are both time- and resource-critical. They are especially resource-critical because the disconnected but still active sub-graph may suffer from buffer over- and under-runs if it cannot be deactivated in time.

One reason is the support of RTP over IP multicast and, therefore, RTP packets from several sources may arrive at the same data port. In this scenario, it is up to the application to react to such a situation. If the decision is, for example, to cache this unexpected data, a new trunk must be created downstream from the RTP receiver. Furthermore, a variable number of clients must be served, if stream scheduling approaches such as patching [40] are implemented. In patching, new receivers must be able to join and leave an ongoing transmission. Since different flows cannot use the same active stream graph concurrently, graph configurations must be changed. The reason for not sharing graphs among several flows is the necessity of keeping per-flow information in SHs, such as encryption keys, file handles, encoding format choices and many more. The implementation of this approach is simpler if one graph per flow is used. Since sockets, semaphores, and SHs with optional threads are supported, several waiting threads exist in each graph. Therefore, this flexible configuration option was chosen.

This requires SHs that can offer endpoints with a variable number of ports that can be opened and closed dynamically. Each endpoint of an SH specifies the number of connections that it can support, and whether or not they can be connected and disconnected dynamically. To allow for fast connection and disconnection of trunks from graphs, it must be possible for the GM to check whether a graph is 'nearly' functional. With the toolkit, each endpoint of an SH can be matched, connected and checked for completeness separately.

C.3.3 Implementor Support

The toolkit is meant to allow the building of a variety of prototype applications for delivery systems that are based on Internet standards. Owing to the interaction of RTP and RTCP and the possibility of receiving data from several sources at a single port, a directed, non-cyclic graph of modules is an appropriate streaming model.

For the most part, packet loss and duplication, delay and jitter occur as usual; but packets from unexpected sources or from peers that should not be sending must also be considered. The toolkit's infrastructure has to support dynamic reconfiguration of the data path, which influences the SH design, as well as the controlling framework, in order to deal with this (sometimes) unexpected behaviour. The ability to reconfigure the graph dynamically is required, for example, for the implementation of a cache: it is necessary to handle user interaction if that cache implements a conditional write-through mode. For example, the client decides to stop watching a video object while the caching process for this object should be continued. In this case, the sub-graph (at the cache) that was used to forward the data to the client can be removed. In reference [184] dynamic reconfiguration of stream graphs with a view to adapting to changing resource availability by reconfiguring SHs is investigated. The requirements for the toolkit presented here are orthogonal to these abilities: in our delivery systems, caches must be able to handle unexpected new streams from the uplink side and pause and continue requests from the client side. To perform this task the GM must split a graph or merge graphs on behalf of the application without disrupting the active data forwarding of a stream.

To allow management of the graphs, a layer of basic classes is given (see Figure C.3) that provides templates and interface definitions for the creation of new SHs. Parent classes with a set of virtual functions ensure the interoperability between SHs. The basic classes are the following:

- *SH*: SH must be inherited by all new SH classes. It provides all interfaces to the GM for configuration, notification and status information.
- *Endpoints*: The endpoint classes provide standard interfaces between the SHs. Each new SH must also include a class that implements its endpoints and inherits from SHEndpoint. SHs can provide both sink (SHSinkEndpoint) and source (SHSourceEndpoint) endpoints.
- *Attributes*: Attributes of a SH are modified by the GM to specialize an SH before it is connected into a graph.
- *Reports*: To implement in-band control, reports are used that provide direct feedback in both directions along the data path. This allows notification between SHs without involvement of the GM. Each endpoint must provide report interfaces that are non-blocking. SHs may communicate via specialized reports even if intermediate SHs cannot interpret them (see section C.3.1).
- *Notifications*: Notifications allow SHs to inform the GM of events, using a non-blocking notification function of the GM. A typical notification concerns the crossing of the threshold in an SH that implements a queue which connects two active sub-graphs. Another use is the notification about RTP packets that arrive from unexpected sources.

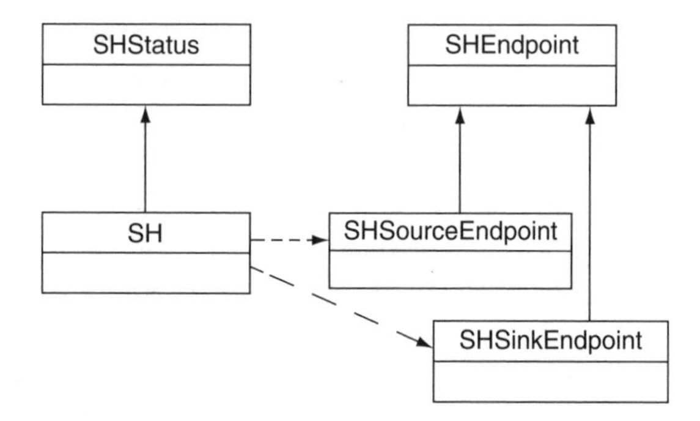

Figure C.3 Parent classes.

C.4 EVALUATION

In this section, the efficiency of the toolkit design is demonstrated by presenting different prototype applications. First, a simple client-server application that is able to simply stream an MPEG-1 movie is shown. Subsequently, the implementation of a cache is presented that is able to either cache data on its local storage or serve a client if the requested object is already in its storage. To be able to measure the efficiency of the toolkit in terms of code reuse the *Frakes and Terry* metric is applied to the cache example. In addition, the implementation of a gleaning capable cache, which shows why reconfiguration is necessary, is presented and the section is concluded with an example on the integration of third-party libraries.

C.4.1 Client-Server Application

An example of the interaction between SHs is the delivery of an MPEG-1 (system stream) movie to a client. Figure C.3 shows the SHs that are used in this simple scenario. The movie is stored on the server's disk. Thus the starting point of the stream path is a *File Source SH*† that reads the data from the disk. In this example, data is requested from the *File Source SH* by the *RTP Encoder SH* which determines the timing in this stream. The *RTP Encoder SH* has knowledge about the actual encoding format of the data and the transport protocol that is used for data transmission. It determines time and amount of data to *pull* from the *File Source SH* and *pushes* it to the *RTP Sink SH* to meet the existing constraints for data rate and delay, and to create a reasonable stream. In the case of an MPEG-1 system stream this means that the *RTP Encoder SH* requests data chunks of equal size and *pushes* those to the *RTP Sink SH*. RTCP receiver reports are interpreted by the *RTP Sink SH* and statistics are forwarded to the *RTP Encoder SH* using the report interface.

The actual stream path is determined by the existing stream graph which represents the layout of the streaming architecture. In Figure C.4 the stream graph at the server consists of GM, *File Source SH*, *RTP Encoder SH* and *RTP Sink SH*. The GM is responsible for the setup and destruction of the SHs, determines the interaction between the individual SHs and represents the interface towards the application. The stream handler functionality of the client is explained in section C.4.5.

† This is described as a source because it is the source of the stream path.

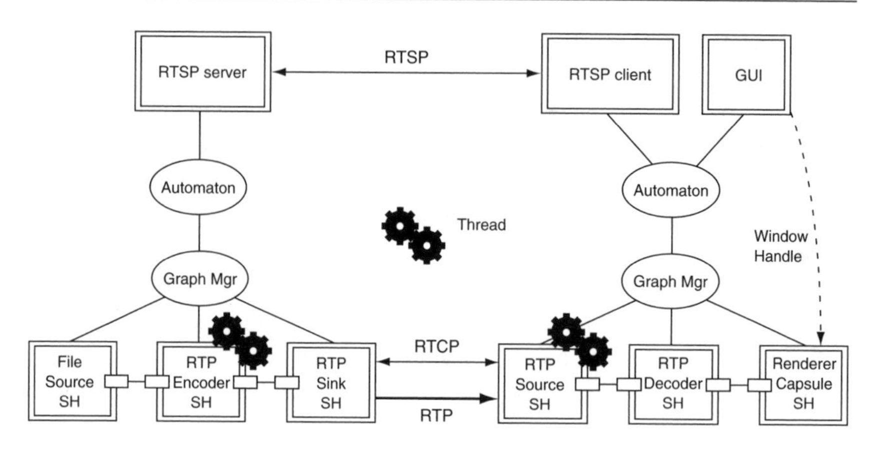

Figure C.4 Client–server configuration overview.

C.4.2 Cache

To enable caching functionality in KOMSSYS only one new SH had to be created. The task of this new SH is to create a copy of the RTP payload and send it on one path to the *RTP Decoder SH* and on the other to the *RTP Sink SH*. An overview of the used SHs and their relation is shown in Figure C.5.

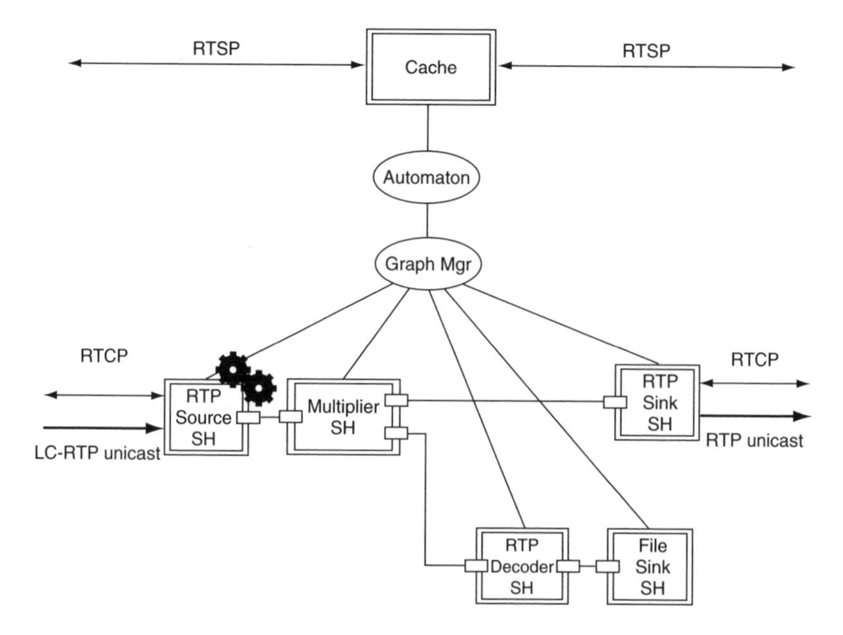

Figure C.5 Cache stream graph.

In the stream graph for the cache, the existing SHs (*FileSink SH*, *RTP Decoder SH* and *RTP Sink SH*) could be used without any modifications. Since the model for the GM is rather static, i.e., a GM cannot be created by, for example, a configuration file but its characteristics are defined by the code that builds the graph manager, a new GM had to be created from scratch. The example of the cache demonstrates how, by the creation of a new SH, caching functionality could be implemented on the stream graph. It must be mentioned that, in order to have the complete functionality of a cache, additional modifications and extensions on the signalling protocol (RTSP) had to be made, but this is not related to the SH subsystem work that is presented here. The code reuse is formally evaluated in section C.4.3.

C.4.3 Reuse

To measure the efficiency of the toolkit in terms of code reuse, two reuse metrics that are described in reference [185] are used, the object-based metric of Banker *et al.* [186], and the 'Frakes and Terry metric with adjustment for complexity' [187], which considers lines of code as well (see Figure C.6). These metrics are chosen because they do not require a development cost factor, which cannot be provided for components of the toolkit. The reuse metrics for the cache, the gleaning capable cache, and the clients are shown in Table C.2. Table C.1 gives an overview of the code statistics of these applications. Since C++ is used, the number of objects differs from the number of classes that are implemented, so values for both objects and classes are provided. The measures used in this work are shown in Figure C.6.

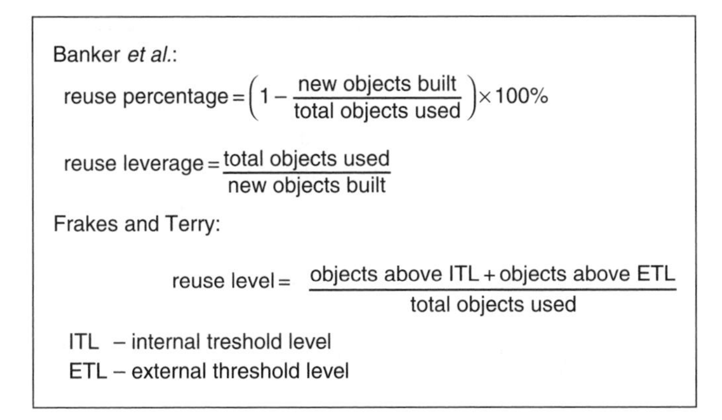

Figure C.6 Reuse measures.

In the measurements presented here, only code on the data path that is composed from SHs and the controlling GM is considered. Other components such as RTSP functionality, which is also necessary to create streaming applications, are not considered.

Table C.1 Code statistics

Application	Existing objects	New objects	Existing classes	New classes	Existing LOC	New LOC
Cache SH pipe only without codecs	4	1	11	3	2271	218
Cache with graph manager and codecs	22	2	36	4	5130	291
Gleaning cache SH pipe only without codecs	6	1	17	3	3506	435
Gleaning cache with graph manager and codecs	13	2	42	4	6365	947
mpeglib support SH pipe only without codecs	3	10	10	6	1690	1082
MPlayer support SH pipe only without codecs	2	1	7	2	2176	273
Gstreamer support SH pipe only without codecs	3	1	10	3	1690	814

The first row of Table C.1 shows the code statistics for the cache example, while the reuse metrics are shown in the first row of Table C.2. As Figure C.5 shows clearly, the graph of SHs that is required for the caching functionality requires only five SHs and their endpoints. For this reason, the reuse level in terms of classes and objects is low, while the high reuse level in lines of code shows that only little additional code was required to build the *Multiplier SH*. The second line considers the existing codecs for RTP payload types, as well as the newly written graph manager for the cache. This increases the number of existing objects considerably but, owing to the size of the graph manager class, the reuse level that considers lines of code is not increased.

Table C.2 Reuse metrics for specific applications

Application	Banker *et al.* reuse leverage and percentage				Frakes and Terry ITL = 0, ETL = 0	
	Objects		Classes		Classes	Lines of code
	leverage	%	leverage	%	reuse level	
Cache SH pipe only without codecs	5	80	4.667	78.57	0.7857	0.9124
Cache with graph manager and codecs	12	91.67	10	90	0.9	0.9463
Gleaning cache SH pipe only without codecs	6	85.71	5.667	85	0.85	0.8896
Gleaning cache with graph manager and codecs	6.5	86.67	10.5	91.3	0.913	0.8705
mpeglib support SH pipe only without codecs	1.3	23.08	2.6	62.5	0.625	0.6097
MPlayer support SH pipe only without codecs	2	66.67	3.5	77.78	0.7778	0.8885
Gstreamer support SH pipe only without codecs	3	75	3.333	76.92	0.7692	0.6749

C.4.4 Gleaning Cache

Reconfiguration plays no role in the example of section C.4.1 but it is a basic requirement for a cache that implements *gleaning*. Roughly, a gleaning cache works by delivering a movie linearly to a client via unicast, which the cache itself receives in two pieces: a short start sequence via unicast and the remaining portion via multicast. For a detailed description of gleaning, the interested reader is referred to reference [4].

A detailed implementation design of a gleaning cache can be found in reference [179]. The cache is not an RTSP proxy as understood in the RFC [62], which caches and redirects only control information. Rather, it is an RTSP/RTP proxy cache that stores content in addition to handling RTSP requests. RTSP messages from different RTSP sessions are multiplexed onto one connection between a server and a cache. RTSP session IDs are the keys to de-multiplex sessions. A cache installs an RTSP connection to a server on demand when a request for the particular server is received from a client. The connection is torn down when no more active RTSP sessions between cache and server exist.

There exist two possible situations that require dynamic reconfiguration of the data path. If the cache does not keep the entire movie, a second client must be served from the same multicast stream and an additional unicast stream, and if the cache keeps the entire movie, the client may decide to pause. In this case, the delivery path to the client must be suspended.

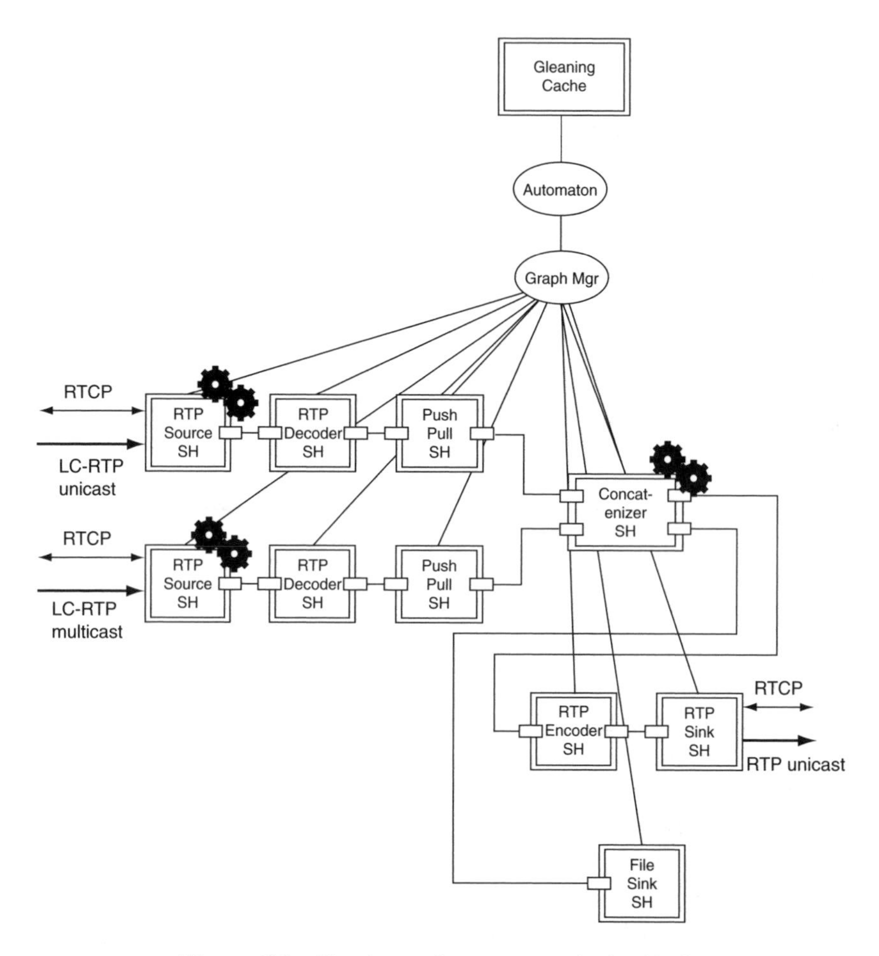

Figure C.7 Gleaning cache stream graphs (caching).

Figures C.7 and C.8 show two possible stream graphs for the gleaning cache with a single client. When a client requests data from the server through the cache, the gleaning cache is often able to join an existing multicast session and request only the missing part of the movie via a unicast stream. When the missing part arrives in a unicast transmission, this data is immediately forwarded to the client while the multicast stream is buffered cyclically and streamed to the client after the unicast stream is finished. If the cache also decides to cache this movie (Figure C.7), both streams are stored linearly on its local disk. If a second client requests the same stream later on, a new graph for playing from the cache is created and reconfiguration is not necessary. In the case where the cache has decided not to cache the title

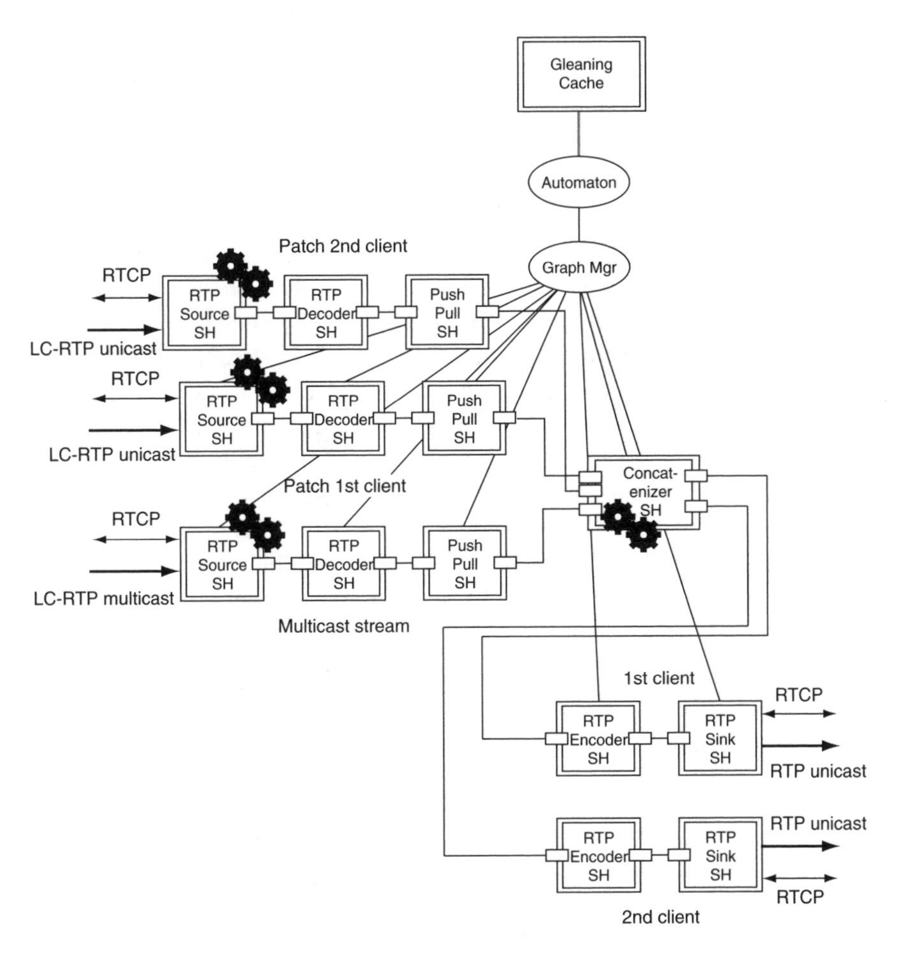

Figure C.8 Gleaning cache stream graphs (two clients).

and the server decides to deliver only a patch stream for this session, the same multicast can be used for the client, but another patch stream has to be requested. The light grey path in Figure C.8 is dynamically created in this case.

If the client pauses and the gleaning cache application decides to continue the caching operation, the trunk of the graph that forwards data to the client must be cut, while the trunk that stores data on disk must be maintained. If the client resumes viewing later, the application must create a new graph, which retrieves the data from the cache. Therefore, two stream paths on the receiving part of the cache are needed: one for the unicast stream and one for the multicast stream. These paths consist of an active *RTP Source SH*, a passive *RTP Decoder SH* and a *PushPull SH*. The latter is needed because the *Concatenizer SH* is active. This is the case because it determines both the order and the timing with which data is forwarded to the client or to the local disk. The main task of the *Concatenizer SH* is to serialize the two incoming streams (multicast and path) and forward them to the client. Data from the multicast stream will be buffered as long as the patch stream is active.

On the data forwarding path to the client, an *RTP Encoder SH* can be seen in through mode, in contrast to its use in active mode in Figure C.3. If active mode is used instead, as in the previous example, the *Concatenizer SH* and the *RTP Encoder SH* would have to be separated by another *PushPull SH*, and both would recreate the required timing of the RTP stream independently.

The two new modules that had to be created from scratch for the data path are *Concatenizer SH* and *Patch GM*. All other SHs that were needed (shown in Figure C.8) to realize the *gleaning* functionality existed already and could be reused in the stream graph. The code statistics and the reuse metrics for the SH and GM are shown in Tables C.1 and C.2, respectively. Although the reuse level in terms of objects is much higher than in the case of the simple cache (see section C.4.2), the reuse level in terms of lines of code is lower.

C.4.5 Third-party Libraries

In this section, an example is given that demonstrates how third-party software can be combined with the toolkit. This approach was used to create several KOMSSYS clients. The variety of decoding and presentation options for MPEG-1 system streams is a good example of the architecture's flexibility. Currently, three alternative tools that can be integrated into stream graphs to perform this task are available with KOMSSYS.

The three tools are the KDE2 library *mpeglib* [188], the stand-alone playback program *mplayer* [189], and the GNOME streaming subsystem *gstreamer* [190]. Figure C.9 depicts the stream graphs that can be set up at the client in the case where it receives a simple unicast MPEG 1 system stream. As in the examples above, *RTP Source SH*, *RTP Decoder SH* and (in case of *mpeglib* and *gstreamer*) *PushPull SH* are reused.

The new SH that had to be built for *mpeglib* is the *MpeglibSink SH*, which works as a wrapper for the library. The *mpeglib* decoder pulls according to the timing that it parses from the arriving MPEG stream. It blocks the SH's thread to maintain this timing.

A similar graph is used for *gstreamer* integration, where an entire open-ended pipe of *gstreamer* SHs is encapsulated by the active *GstMpegDemux SH*, and the SH sink endpoint implements a *gstreamer* sink pad. In this

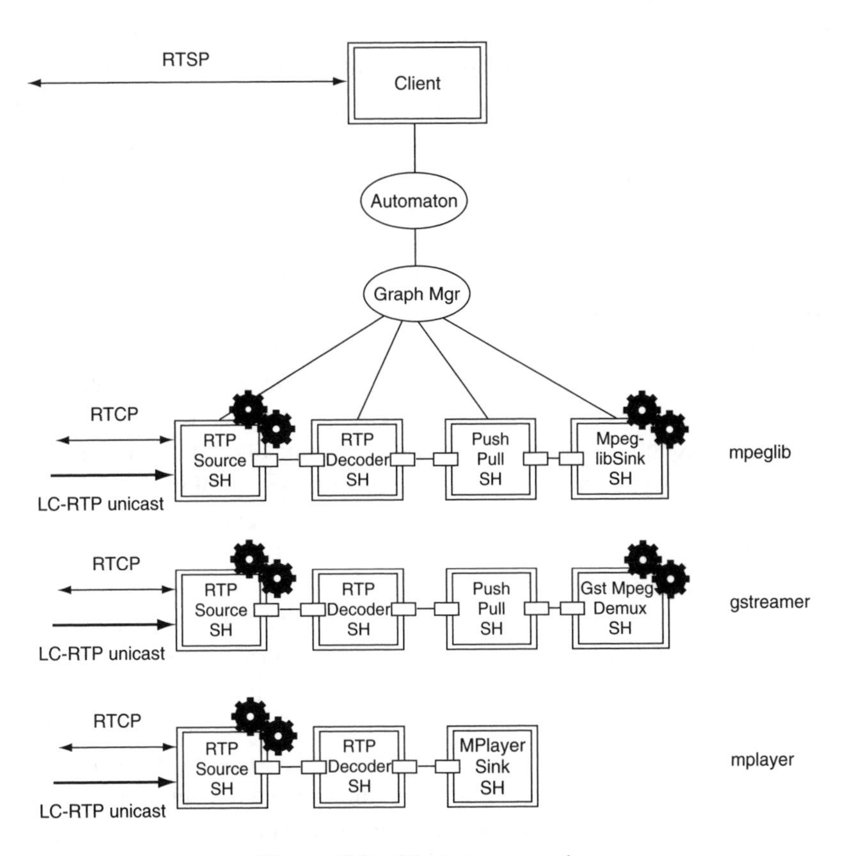

Figure C.9 Client stream graph.

case, *gstreamer* comes with a co-thread package that is not compatible with POSIX threads and, therefore, must be encapsulated and hidden by the SH.

In the third approach, the stand-alone executable *mplayer* is encapsulated by the *MPlayerSink SH*. The SH forks to start the player, and forwards data to it through a Unix socket. Since this works in push mode, the *PushPull SH* is not required in this configuration. Thus by simply creating a different stream graph, three different MPEG-1 decoding libraries can be used in the client.

Code statistics and reuse levels of the three SHs can be found in Tables C.1 and C.2.

C.5 SUMMARY

Over the years the KOMSSYS platform has been constantly extended to develop research prototypes in the area of wide-area distribution systems for streaming media in the Internet. The platform could be considered as an appropriate abstraction for developing streaming applications, especially since a lot of recent implementation work has been done by developers [191, 192, 179] who were not involved in the actual SH design. Based on the results presented in section C.4.3, one can claim that KOMSSYS is easy to extend, for newcomers as well as for the original developers.

Two special requirements are needed to perform investigations in VoD distribution infrastructures: the ability to integrate third-party software and the ability to share resources among user-initiated sessions. The first requirement prevented, for example, the use of co-routines and the limitation to a single mechanism for awaiting events. The second requirement led to the intention of reconfiguring the stream graph dynamically. The resulting toolkit is most similar to the InfoPipe [171] design, although InfoPipe does not support dynamic reconfiguration of the data path. One central difference is that InfoPipe components use a special co-routine model while, with the toolkit used for KOMSSYS, native threads are used. The presented approach simplifies the integration of third-party libraries and tools. The InfoPipe approach is more efficient, but it requires conformance to a programming model.

In KOMSSYS, the initial code base considered mainly the distribution of a CBR MPEG-1 system and MP3 streams in a caching hierarchy and to receiving clients. From the start of the development, the integration of third-party software was an important element, for example, alternative streaming subsystems are the commercial servers, VideoCharger and RealServer. The initial formats were chosen because they combine hardware- and OS-independent playback capability with an appropriate quality. More recently, H.261 and VBR MPEG-I audio and SPEG, video were added.

The need for dynamic data paths was not an original requirement. The intention of building a gleaning prototype led to a redesign, which resulted in the decisions that are presented in section C.4. An important observation is that the new design does not increase the number of context changes or the buffer requirements for the data path if one of the simple data paths is configured. On the contrary, the attachable threads extension allows SHs to get rid of some threads that were not exposed to the interface before. The reason is that attachable threads can be shared between SHs that are separated by a sub-graph that must use an incompatible mechanism to wait for events. An example is the RTCP thread of the RTP sink SH, which appears as a passive SH to the GM. Its thread, not exposed to the GM, is responsible for receiving and processing control messages.

Although the applicability and extensibility of the approach has been shown with the gleaning cache prototype, better handling, by integrating the SH approach consequently into the client, was also achieved.

References

[1] R. Steinmetz and K. Nahrstedt. *Media Coding and Content Processing*. Prentice Hall PTR, 2002. ISBN 0-13-031399-8.

[2] S. McCreary and K. Claffy. Trends in Wide Area IP Traffic Patterns. In *Proceedings of 13th ITC Specialist Seminar on Internet Traffic Measurement and Modeling*, September 2000. http://www.caida.org/outreach/papers/AIX0005.

[3] T. D. Little and D. Venkatesh. Prospects for Interactive Video-on-Demand. *IEEE Multimedia*, **1**(3):14–25, May 1994.

[4] C. Griwodz. *Wide-area True Video-on-Demand by a Decentralized Cache-based Distribution Infrastructure*. PhD thesis, Darmstadt University of Technology, Darmstadt, Germany, June 2000.

[5] C. Griwodz, M. Bär and L. C. Wolf. Long-term Movie Popularity in Video-on-Demand Systems. In *Proceedings of ACM Multimedia Conference 1997, Seattle, WA, USA*, pages 340–357, November 1997.

[6] R. Braden, D. Clark and S. Shenker. RFC 1633 – Integrated Services in the Internet Architecture: an Overview. Informational Track RFC, June 1994.

[7] J.-Y. Lee, T.-H. Kim and S.-J. Ko. Motion Prediction Based on Temporal Layering for Layered Video Coding. In *Proceedings ITC-CSCC'98*, pages 245–248, July 1998.

[8] K. Shen and E. J. Delp. Wavelet Based Rate Scalable Video Compression. *IEEE Transactions on Circuits and Systems for Video Technology*, **9**(1):109–122, February 1999.

[9] A. S. Tanenbaum and M. van Steen. *Distributed Systems*. Prentice Hall, 2002. ISBN 0-13-088893-1.

[10] M. Satyanarayanan, J. Kistler, P. Kumar, M. Okasaki, E. Siegel and D. Steere. Coda: A Highly Available File System for a Distributed Workstation Environment. *IEEE Transactions on Computers*, **39**(4):447–459, April 1990.

[11] I. Foster and C. Kesselman. *The Grid*. Morgan Kaufmann Publishers, San Francisco, CA, USA, 1998. ISBN 1-55860-475-8.

[12] International Organization for Standardization (ISO). *Information Technology – Coding of Moving Pictures and Associated Audio for Digital Storage Media at up to about 1.5 Mbit/s* – Part 3: *Audio*. International Standard, 1996. ISO/IEC 11172-3:1993/Cor 1:1996.

Scalable Video on Demand: Adaptive Internet-based Distribution M. Zink

[13] G. On, J. B. Schmitt and R. Steinmetz. Quality of Availability: Replica Placement for Widely Distributed Systems. In *Proceedings of the 11th IEEE/IFIP International Workshop on Quality of Service (IWQoS'03), Monterey, CA, USA*, pages 325–342, June 2003. ISBN 3-540-40281-0.

[14] International Organization for Standardization (ISO). *Coding of Moving Pictures and Audio, MPEG-21 Overview*. International Standard, 2002. ISO/IEC JTC1/SC29/WG11/N4801.

[15] D. Wessels and K. Claffy. RFC 2186 – Internet Cache Protocol (ICP), version 2. Informational Track RFC, September 1997.

[16] H. Schulzrinne, A. Rao and R. Lanphier. RFC 2326 – Real Time Streaming Protocol (RTSP). Standards Track RFC, April 1998.

[17] C. Küfner. *Möglichkeiten für den Einsatz von Lastverteilungsstrategien verteilter Systeme in der Videoverteilung* (in German). Studienarbeit. Fachbereich Elektrotechnik und Informationstechnik, TU-Darmstadt, September 1998.

[18] R. Fielding, J. Gettys, J. Mogul, H. Frystyk, L. Masinter, P. Leach and T. Berners-Lee. RFC 2616 – Hypertext Transfer Protocol – HTTP/1.1. Standards Track RFC, 1999.

[19] R. Tewari. *Architectures and Algorithms for Scalable Wide-area Information Systems*. PhD thesis, University of Texas, Austin, TX, USA, August 1998.

[20] International Organization for Standardization (ISO). *Information Technology – Coding of Moving Pictures and Associated Audio for Digital Storage Media at up to about 1.5 Mbit/s* – Part 1: *Systems*. International Standard, 1993. ISO/IEC 11172-1:1993.

[21] J. Almeida, J. Krueger, D. L. Eager and M. K. Vernon. Analysis of Educational Media Server Workloads. In *Proceedings of the 11th Annual Workshop on Network and Operating System Support for Digital Audio and Video, Port Jefferson, NY, USA*, pages 21–30, June 2001. ISBN 1-58113-370-7.

[22] R. Braden, L. Zhang, S. Berson, S. Herzog and S. Jamin. RFC 2205 – Resource ReSerVation Protocol (RSVP) – Version 1 Functional Specification. Standards Track RFC, September 1997.

[23] J. B. Schmitt. *Heterogeneous Network Quality of Service Systems*. Klu-wer Academic Publishers, Dordrecht, The Netherlands, 2001. ISBN 0-7923-7410-X.

[24] S. D. Gribble and E. A. Brewer. System Design Issues for Internet Mid-dleware Services: Deductions from a Large Client Trace. In *Proceedings of the USENIX Symposium on Internet Technologies and Systems, Monterey, CA, USA*, December 1997.

[25] A. Wolman, G. Voelker, N. Sharma, N. Caldwell, M. Brown, T. Landray, D. Pinnel, A. Karlin and H. Levy. Organization-based Analysis of Web-object Sharing and Caching. In *Proceedings of USITS'99: The 2nd USENIX Symposium on Internet Technologies and Systems, Boulder, CO, USA*, October 1999.

[26] M. Chesire, A. Wolman, G. Voelker and H. Levy. Measurement and Analysis of a Streaming-Media Workload. In *Proceedings of USITS'02: The 3rd USENIX Symposium on Internet Technologies and Systems, San Francisco, CA, USA*, March 2001.

[27] Streaming Media Growth Analysis & Market Dynamics: 2002 vs. 2001, Executive Summary. http://www.streamingmedia.com/Research.

[28] J. Widmer, R. Denda and M. Mauve. A Survey on TCP-Friendly Congestion Control. *Special Issue of the IEEE Network Magazine 'Control of Best Effort Traffic'*, **15**(3):28–37, May 2001.

[29] S. Acharya and B. Smith. Experiment to Characterize Videos Stored on the Web. In *Proceedings of SPIE/ACM Conference on Multimedia Computing and Networking (MMCN), San Jose, CA, USA*, pages 166–178, January 1998.

[30] S. Acharya, B. Smith and P. Parnes. Characterizing User Access to Videos on the World Wide Web. In *Proceedings of SPIE/ACM Conference on Multimedia Computing and Networking (MMCN), San Jose, CA, USA*, pages 130–141. SPIE, January 2000.

[31] D. Sitaram and A. Dan. *Multimedia Servers*. Morgan Kaufmann Publishers, 2000. ISBN 1-55860-430-8.

[32] A. Hu. Video-on-Demand Broadcasting Protocols: a Comprehensive Study. In *Proceedings of the 20th Annual Joint Conference of the IEEE Computer and Communications Societies (INFOCOM'01), Anchorage, AK, USA*, pages 508–517. IEEE Computer Society Press, April 2001.

[33] K. A. Hua and S. Sheu. Skyscraper Broadcasting: a New Broadcasting Scheme for Metropolitan Video-on-Demand Systems. In *Proceedings of the ACM SIGCOMM '97, Cannes, France*, pages 89–100, September 1997.

[34] J.-F. Paris, D. D. E. Long and P. E. Mantey. Zero-Delay Broadcasting Protocols for Video-on-Demand. In *Proceedings of the ACM Multime-dia Conference 1999, Orlando, FL, USA*, pages 189–197, November 1999.

[35] D. Eager, M. Vernon and J. Zahorjan. Minimizing Bandwidth Requirements for On-Demand Data Delivery. *IEEE Transactions on Knowledge and Data Engineering*, **13**(5):742–757, 2001.

[36] J.-P. Nussbaumer, B. Patel, F. Schaffa and J. P. G. Sterbenz. Networking Requirements for Interactive Video on Demand. *IEEE Journal on Selected Areas in Communications*, **13**(5):779–787, June 1995. ISSN 0733-8716.

[37] G. Bianchi and R. Melen. Non Stationary Request Distribution in Video-on-Demand Networks. In *Proceedings of the 16th Annual Joint Conference of the IEEE Computer and Communications Societies (INFOCOM'97), Kobe, Japan*, pages 711–717. IEEE Computer Society Press, April 1997.

[38] C. Griwodz, O. Merkel, J. Dittmann and R. Steinmetz. Protecting VoD the Easier Way. In *Proceedings of ACM Multimedia Conference 1998, Bristol, UK*, pages 21–28, September 1998. ISBN 0-201-30990-4.

[39] J. M. Boyce and R. D. Gaglianello. Packet Loss Effects on MPEG Video Sent over the Public Internet. In *Proceedings of ACM Multimedia Conference 1998, Bristol, UK*, pages 181–190, September 1998. ISBN 0-201-30990-4.

[40] K. A. Hua, Y. Cai and S. Sheu. Patching: a Multicast Technique for True Video-on-Demand Services. In *Proceedings of the ACM Multime-dia Conference 1998, Bristol, UK*, pages 191–200, September 1998. ISBN 0-201-30990-4.

[41] D. L. Eager, M. K. Vernon and J. Zahorjan. Bandwidth Skimming: a Technique for Cost-effective Video-on-Demand. In *Proceedings of SPIE/ACM Conference on Multimedia Computing and Networking (MMCN), San Jose, CA, USA*, pages 206–215. SPIE, January 2000.

[42] R. Frederick, J. Geagan, M. Kellner and A. Periyannan. Caching Support in RTSP/RTP Servers. Internet Draft, March 2000. Work in Progress.

[43] Y. Wang, M. Claypool and Z. Zhu. An Empirical Study of Realvideo Performance across the Internet. In *First ACM SIGCOMM Workshop on Internet Measurement Workshop, San Francisco, CA, USA*, pages 295–309, November 2001.

[44] J. Chung, M. Claypool and Y. Zhu. Measurement of the Congestion Responsiveness of Realplayer Streaming Video over UDP. In *Packet Video Workshop 2003, Nantes, France*, April 2003.

[45] K. Rao, Z. S. Bojkovic and D. A. Milanovic. *Multimedia Communication Systems*. Prentice Hall, 2002. ISBN 0-13-031398-X.

[46] R. Rejaie, M. Handley and D. Estrin. RAP: an End-to-End Rate-based Congestion Control Mechanism for Realtime Streams in the Internet. In *Proceedings of the Eighteenth Annual Joint Conference of the IEEE Computer and Communications Societies 1999 (INFOCOM'99), New York, NY, USA*, pages 395–399, March 1999.

[47] S. Floyd, M. Handley, J. Padhye and J. Widmer. Equation-Based Congestion Control for Unicast Applications. In *Proceedings of the ACM SIGCOMM'00 Conference on Applications, Technologies, Architectures, and Protocols for Computer Communication 2000, Stockholm, Sweden*, pages 43–56, August 2000.

[48] W.-T. Tan and A. Zakhor. Real-time Internet Video Using Error Resilient Scalable Compression and TCP-friendly Transport Protocol. *IEEE Transactions on Multimedia*, **1**(2):172–186, 1999.

[49] Universitärer Lehrverbund Informatik. http://www.uli-campus.de/index_en.html.

[50] MIT Open Courseware. http://ocw.mit.edu.

[51] RealOne Player. http://www.real.com.

[52] Technical White Paper: Video Content Distribution, Kasenna Inc. http://www.kasenna.com.

[53] Network Appliance Delivers Global Video Streaming Solution, Network Appliance Inc. http://www.netapp.com/case_studies/streaming.html.

[54] Inktomi Traffic Server – Media Cache Option, Inktomi Inc. http://www.inktomi.com/products/cns/resources/technical.html.

[55] High Performance Proxy Caching with Blue Coat, Blue Coat Inc. http://www.bluecoat.com/solutions/proxy_server.html.

[56] Media Servers and Media Proxies – the Critical Differences, Infolibria Inc. http://www.infolibria.com/products/collateral/Application_Briefs/ds_ab_strea%m_media_v7e.pdf.

[57] Novell's Volera Media Ecelerator. http://www.novell.com/products/volera/media.html.

[58] Certeon's Media Mall. http://www.certeon.com/products_and_services/mediamall_overview.html.

[59] S. Sen, J. Rexford and D. Towsley. Proxy Prefix Caching for Multimedia Streams. In *Proceedings of the Eighteenth Annual Joint Conference of the IEEE Computer and Communications Societies 1999 (INFO-COM'99), New York, NY, USA*, pages 1310–1319, March 1999.

[60] Akamai. http://www.akamai.com.

[61] Cable & Wireless. http://www.cw.com.

[62] H. Schulzrinne, S. L. Casner, R. Frederick and V. Jacobson. RFC 1889 – RTP: a Transport Protocol for Real-Time Applications. Standards Track RFC, January 1996.

[63] H. Schulzrinne. RFC 1890 – RTP Profile for Audio and Video Conferences with Minimal Control. Standards Track RFC, January 1996.

[64] H. Schulzrinne, S. L. Casner, R. Frederick and V. Jacobson. RTP: a Transport Protocol for Real-Time Applications. Internet Draft, March 2003. Work in Progress.

[65] M. Handley and V. Jacobson. RFC 2327 – SDP: Session Description Protocol. Standards Track RFC, April 1998.

[66] H. Schulzrinne, A. Rao and R. Lanphier. RFC 2326 – Real Time Streaming Protocol (RTSP). Standards Track RFC, April 1998.

[67] D. Kutscher, J. Ott and C. Bormann. Session Description and Capability Negotiation. Internet Draft, July 2002. Work in Progress.

[68] M. Westerlund and T. Zeng. Extended RTP Profile for RTCP-based Feedback. Internet Draft, June 2003. Work in Progress.

[69] U. Roedig. *Firewall Architectures for Multimedia Applications* (in German). PhD thesis, Darmstadt University of Technology, Darmstadt, Germany, November 2002.

[70] E. Kohler, M. Handley, S. Floyd and J. Padhye. Datagram Congestion Control Protocol (DCCP). Internet Draft, October 2002. Work in Progress.

[71] S. Floyd, E. Kohler and J. Padhye. Profile for DCCP Congestion Control ID 3: TFRC Congestion Control. Internet Draft, March 2003. Work in Progress.

[72] R. R. Stewart, Q. Xie, K. Morneault, C. Sharp, H. J. Schwarzbauer, T. Taylor, I. Rytina, M. Kalla, L. Zhang and V. Paxson. RFC 2960 – Stream Control Transmission Protocol. Standards Track RFC, October 2000.

[73] M. S. Day, B. Cain, G. Tomlinson and P. Rzewski. RFC 3466 – A Model for Content Internetworking (CDI). Standards Track RFC, 2003.

[74] A. Mauthe. Content Management and Delivery – Related Technology Areas. Technical Report MPG-03-04, Lancaster University, October 2002.

[75] Kingston Communications. http://www.kingston-comms.com.

[76] R. Frederick. RTSP Interoperability Bakeoff . In *Proceedings of the Forty-eighth Internet Engineering Task Force Meeting, Pittsburgh, PA, USA*, August 2000.

[77] F. Pereira and T. Ebrahimi. *The MPEG-4 Book*. Prentice Hall, 2002. ISBN 0-13-061621-4.

[78] E. Amir, S. McCanne and M. Vetterli. A Layered DCT Coder for Inter-net Video. In *International Conference on Image Processing, Lau-sanne, Switzerland*, pages 13–16, September 1996.

[79] J. M. Shapiro. Embedded Image Coding Using Zerotrees of Wavelet Coefficients. *IEEE Transactions on Signal Processing*, **41**(12):3445–3462, December 1993.

[80] J.-Y. Lee. *A Novel Approach to Multi-Layer Coding of Video for Playback Scalability*. PhD thesis, Graduate School Korea University, Seoul, Korea, June 1999.

[81] C. Krasic and J. Walpole. Priority-Progress Streaming for Quality-Adaptive Multimedia. In *ACM Multimedia Doctoral Symposium, Ottawa, Canada*, pages 463–464, October 2001.

[82] International Organization for Standardization (ISO). *Generic Coding of Moving Pictures and Associated Audio Information* – Part2: *Video*. International Standard, 2000. ISO/IEC 13818-2:2000.

[83] ITU-T. *Video Coding for Low Bit Rate Communication*. International Standard, 1995. ITU-T Recommendation H.263.

[84] W. Li. Overview of Fine Granularity Scalability in MPEG-4 Video Standard. *IEEE Transactions on Circuits and Systems for Video Technologies*, **11**(3):301–317, March 2001.

[85] A. R. Reibman, H. Jafarkhani, Y. Wang, M. T. Orchard and R. Puri. Multiple Description Video Coding Using Motion-compensated Temporal Prediction. *IEEE Transactions on Circuits and Systems for Video Technology*, **12**(3):193–204, March 2002.

[86] V. K. Goyal. Multiple Description Coding: Compression Meets the Network. *IEEE Signal Processing Magazine*, **18**(5):74–93, September 2001.

[87] J. Apostolopoulos, T. Wong, S. Wee and D. Tan. On Multiple Description Streaming with Content Delivery Networks. In *Proceedings of the 21st Annual Joint Conference of the IEEE Computer and Communications Societies (INFOCOM'02), New York, NY, USA*, pages 1736–1745, June 2002.

[88] Y.-C. Lee, J. Kim, Y. Altunbasak and R. Mersereau. Performance Comparison of Layered and Multiple Description Coded Video Streaming over Error-prone Networks. In *Proceedings of the International Conference on Communications, Anchorage, AK, USA*, pages 35–39, May 2003.

[89] D. Loguinov and H. Radha. End-to-End Internet Video Traffic Dynamics: Statistical Study and Analysis. In *Proceedings of the 21st Annual Joint Conference of the IEEE Computer and Communications Societies (INFOCOM'02)*, pages 723–732, June 2002.

[90] I. Rhee, N. Balaguru and G. Rouskas. Mtcp: Scalable TCP-like Congestion Control for Reliable Multicast. In *Proceedings of the 19th Annual Joint Conference of the IEEE Computer and Communications Societies (INFOCOM'99), New York, NY, USA*, pages 1265–1273, March 1999.

[91] L. Rizzo. pgmcc: a TCP-friendly Single-rate Multicast Congestion Control Scheme. In *ACM SIGCOMM '00 Conference on Applications, Technologies, Architectures, and Protocols for Computer Communication, Stockholm, Sweden*, pages 17–28, August 2000.

[92] M. Zink, C. Griwodz, J. Schmitt and R. Steinmetz. Exploiting the Fair Share to Smoothly Transport Layered Encoded Video into Proxy Caches. In *Proceedings of SPIE/ACM Conference on Multimedia Computing and Networking (MMCN), San Jose, CA, USA*, pages 61–72. SPIE, January 2002. ISBN 0-8194-4413-8.

[93] J. Widmer and M. Handley. Extending Equation-Based Congestion Control to Multicast Applications. In *Proceedings of the ACM SIG-COMM '01 Conference on Applications, Technologies, Architectures, and Protocols for Computer Communication 2001, San Diego, CA, USA*, pages 275–285, August 2001.

[94] D. Bansal and H. Balakrishnan. Binomial Congestion Control Algorithms. In *Proceedings of the 20th Annual Joint Conference of the IEEE Computer and Communications Societies (INFOCOM'01), Anchorage, AK, USA*, pages 631–640, April 2001.

[95] D. Bansal, H. Balakrishnan, S. Floyd and S. Shenker. Dynamic Behavior of Slowly-Responsive Congestion Control Algorithms. In *Proceedings of the ACM SIGCOMM '01 Conference on Applications, Technologies, Architectures, and Protocols for Computer Communication 2001, San Diego, CA, USA*, pages 263–274, August 2001.

[96] R. Rejaie, M. Handley and D. Estrin. Quality Adaptation for Congestion Controlled Video Playback over the Internet. In *Proceedings of the ACM SIGCOMM '99 Conference on Applications, Technologies, Architectures, and Protocols for Computer Communication 1999, New York, NY, USA*, pages 189–200, August 1999.

[97] S. Nelakuditi, R. R. Harinath, E. Kusmierek and Z.-L. Zhang. Providing Smoother Quality Layered Video Stream. In *Proceedings of the 10th International Workshop on Network and Operating System Support for Digital Audio and Video (NOSSDAV), Chapel Hill, NC, USA*, June 2000.

[98] N. Feamster, D. Bansal and H. Balakrishnan. On the Interactions Between Layered Quality Adaptation and Congestion Control for Streaming Video. In *11th International Packet Video Workshop (PV2001), Kyongju, Korea*, April 2001.

[99] D. Saparilla and K. W. Ross. Optimal Streaming of Layered Video. In *Proceedings of the Nineteenth Annual Joint Conference of the IEEE Computer and Communications Societies 2000 (INFOCOM'00), Tel-Aviv, Israel*, pages 737–746, March 2000.

[100] S. McCanne, M. Vetterli and V. Jacobson. Receiver-driven Layered Multicast. In *Proceedings of the ACM SIGCOMM'96 Conference on Applications, Technologies, Architectures, and Protocols for Computer Communication 1996, Palo Alto, CA, USA*, pages 117–130, August 1996.

[101] R. Gopalakrishnan, J. Griffioen, G. Hjalmtysson, C. Sreenan and S. Wen. A Simple Loss Differentiation Approach to Layered Multicast. In *Proceedings of the Nineteenth Annual Joint Conference of the IEEE Computer and Communications Societies 2000 (INFOCOM'00), Tel-Aviv, Israel*, pages 461–469, March 2000.

[102] R. Tewari, H. Vin, A. Dan and D. Sitaram. Resource-Based Caching for Web Servers. In *Proceedings of SPIE/ACM Conference on Multimedia Computing and Networking (MMCN), San Jose, CA, USA*, pages 191–204, January 1998.

[103] K.-L. Wu, P. S. Yu and J. L. Wolf. Segment-Based Proxy Caching of Multimedia Streams. In *Proceedings of the Tenth International World Wide Web Conference, Hong Kong, China*, pages 36–44, May 2001.

[104] E. Balafoutis, A. Panagakis, N. Laoutaris and I. Stavrakakis. The Impact of Replacement Granularity on Video Caching. In *Proceedings of the 2nd IFIP-TC6 Networking Conference, Pisa, Italy*, pages 214–225. IEEE/IFIP, May 2002.

[105] S.-H. G. Chan and F. Tobagi. Distributed Server Architectures for Networked Video Services. *IEEE/ACM Transactions on Networking*, **9**(2):125–136, April 2001.

[106] Z. Miao and A. Ortega. Proxy Caching for Efficient Video Services over the Internet. In *Proceedings of the 9th Packet Video Workshop, New York, NY, USA*, pages 36–44, April 1999.

[107] S. Ramesh, I. Rhee and K. Guo. Multicast with Cache (Mcache): an Adaptive Zero Delay Video-on-Demand Service. In *Proceedings of the 20th Annual Joint Conference of the IEEE Computer and Communications Societies (INFOCOM'01), Anchorage, AK, USA*, pages 85–94. IEEE Computer Society Press, April 2001.

[108] B. Wang, S. Sen, M. Adler and D. Towsley. Optimal Proxy Cache Allocation for Efficient Streaming Media Distribution. In *Proceedings of the 21st Annual Joint Conference of the IEEE Computer and Communications Societies (INFOCOM'02), New York, NY, USA*, pages 1726–1735, June 2002.

[109] M. Hofmann, T. S. E. Ng, K. Guo, P. Sanjoy and H. Zhang. Caching Techniques for Streaming Multimedia over the Internet. Technical report, Bell Labs, Holmdel, NJ, USA, May 1999.

[110] J. Rexford, S. Sen and A. Basso. A Smoothing Proxy Service for Variable-bit-rate Streaming Video. In *Proceedings of Global Internet Symposium, Rio de Janeiro, Brazil*, pages 1823–1829, December 1999.

[111] O. Verscheure, C. Venkatramani, P. Frossard and L. Amini. Joint Server Scheduling and Proxy Caching for Video Delivery. *Computer Communications Journal, Elsevier Science*, **25**(4):413–423, March 2002.

[112] D. L. Eager, M. C. Ferris and M. K. Vernon. Optimized Caching in Systems with Heterogeneous Client Populations. *Performance Evaluation*, **42**(2/3):163–185, 2000.

[113] J. Almeida, D. L. Eager, M. Ferris and M. K. Vernon. Provisioning Content Distribution Networks for Streaming Media. In *Proceedings of the 21st Annual Joint*

Conference of the IEEE Computer and Communications Societies (INFOCOM'02), New York, NY, USA, pages 1746–1755. IEEE Computer Society Press, June 2002.

[114] Z.-L. Zhang, Y. Wang, D. H. Du and D. Su. Prospects for Interactive Video-on-Demand. *IEEE/ACM Transactions on Networking*, **8**(4):429–442, August 2000.

[115] R. Rejaie, H. Yu, M. Handley and D. Estrin. Multimedia Proxy Caching for Quality Adaptive Streaming Applications in the Internet. In *Proceedings of the Nineteenth Annual Joint Conference of the IEEE Computer and Communications Societies 2000 (INFOCOM'00), Tel-Aviv, Israel*, pages 980–989, March 2000.

[116] S. Paknikar, M. Kankanhalli, K. Ramakrishnan, S. Srinivasan and L. H. Ngoh. A Caching and Streaming Framework for Multimedia. In *Proceedings of the ACM Multimedia Conference 2000, Los Angeles, CA, USA*, pages 13–20, October 2000.

[117] J. Kangasharju, F. Hartanto, M. Reisslein and K. W. Ross. Distributing Layered Encoded Video through Caches. *IEEE Transactions on Computers*, **51**(6):622–636, June 2002.

[118] R. Rejaie and J. Kangasharju. Mocha: a Quality Adaptive Multimedia Proxy Cache for Internet Streaming. In *Proceedings of the 11th International Workshop on Network and Operating System Support for Digital Audio and Video (NOSSDAV'01), Port Jefferson, NY, USA*, pages 3–10, June 2001.

[119] Squid Web Proxy Cache. http://www.squid-cache.org/.

[120] J. Liu, X. Chu and J. Xu. Proxy Cache Management for Fine-Grained Scalable Video Streaming. In *Proceedings of the 23rd Annual Joint Conference of the IEEE Computer and Communications Societies (IN-FOCOM'04), New York, NY, USA*, pages 1490–1500, March 2004.

[121] S. Floyd, V. Jacobson, C. Liu, S. McCanne and L. Zhang. A Reliable Multicast Framework for Light-weight Sessions and Application Level Framing. *IEEE/ACM Transactions on Networking*, **5**(6):784–803, 1997.

[122] B. Sabata, M. J. Brown, B. A. Denny and Chung-ho Heo. Transport protocol for reliable multicast: TRM. In *Proceedings of the IASTED International Conference on Networks, Orlando, FL*, pages 143–145, January 1996.

[123] S. Paul, K. K. Sabnani, J. C. Lin and S. Bhattacharyya. Reliable Multicast Transport Protocol (RMTP). *IEEE Journal on Selected Areas in Communications*, **15**(3):407–421, 1997.

[124] T. Liao. Light-weight Reliable Multicast Protocol. Technical report, In-ria, France, August 1998.

[125] J. Nonnenmacher, E. Biersack and D. Towsley. Parity-based Loss Recovery for Reliable Multicast Transmission. In *Proceedings of the ACM SIGCOMM '97 Conference on Applications, Technologies, Architectures, and Protocols for Computer Communication, Cannes, France*, pages 289–300, September 1997.

[126] N. Feamster and H. Balakrishnan. Packet Loss Recovery for Streaming Video. In *12th International Packet Video Workshop (PV2002), Pitts-burgh, PA, USA*, April 2002.

[127] M. Zink, C. Griwodz, A. Jonas, and R. Steinmetz. LC-RTP (Loss Collection RTP): Reliability for Video Caching in the Internet. In *Proceedings of the Seventh International Conference on Parallel and Distributed Systems: Workshops, Iwate, Japan*, pages 281–286. IEEE, July 2000. ISBN 0-7695-0571-6.

[128] S. Acharya and B. Smith. MiddleMan: a Video Caching Proxy Server. In *Proceedings of NOSSDAV 2000, Chapel Hill, NC, USA*, June 2000.

[129] Y. Chae, K. Guo, M. M. Buddhikot, S. Suri and E. W. Zegura. Silo, Rainbow, and Caching Token: Schemes for Scalable, Fault Tolerant Stream Caching. *IEEE Journal on Selected Areas in Communications*, **20**(7):1328–1344, 2002.

[130] M. Castro, P. Druschel, Y. C. Hu and A. Rowstron. SplitStream: High-bandwidth Content Distribution in Cooperative Environments. In *Proceedings of the 2nd International Workshop on Peer-to-Peer Systems (IPTPS '03), Berkeley, CA, USA*, pages 103–107, February 2003.

[131] C. Krasic and J. Walpole. QoS Scalability for Streamed Media Delivery. Technical Report OGI CSE Technical Report CSE-99-011, Oregon Graduate Institute of Science & Technology, September 1999.

[132] J. Lu. Signal Processing for Internet Video Streaming: a Review. In *Proceedings of SPIE Image and Video Communications and Processing, San Jose, CA, USA*, pages 246–259. SPIE, January 2000.

[133] M. Zink, J. Schmitt and R. Steinmetz. Retransmission Scheduling in Layered Video Caches. In *Proceedings of the International Conference on Communications 2002 (ICC'02), New York, NY, USA*, pages 2474–2478, April 2002. ISBN 0-7803-7401-0.

[134] T. Alpert and J.-P. Evain. Subjective quality evaluation – the SSCQE and DSCQE methodologies. EBU Technical Review, February 1997.

[135] R. Aldridge, D. Hands, D. Pearson and N. Lodge. Continuous Quality Assessment of Digitally-coded Television Pictures. *IEE Proceedings on Vision, Image and Signal Processing*, **145**(2):116–123, 1998.

[136] ITU-R: Methodology for the Subjective Assessment of the Quality of Television Picture. International Standard, 2000. ITU-R BT.500-10.

[137] R. Aldridge, J. Davidoff, M. Ghanbari, D. Hands and D. Pearson. Measurement of Scene-dependent Quality Variations in Digitally Coded Television Pictures. *IEE Proceedings on Vision, Image and Signal Processing*, **142**(3):149–154, 1995.

[138] F. Pereira and T. Alpert. MPEG-4 Video Subjective Test Procedures and Results. *IEEE Transactions on Circuits and Systems for Video Technology*, **7**(1):32–51, 1997.

[139] M. Masry and S. Hemami. An Analysis of Subjective Quality in Low Bit Rate Video. In *International Conference on Image Processing (ICIP), 2001, Thessaloniki, Greece*, pages 465–468. IEEE Computer Society Press, October 2001.

[140] C. Kuhmünch and C. Schremmer. Empirical Evaluation of Layered Video Coding Schemes. In *Proceedings of the IEEE International Conference on Image Processing (ICIP), Thessaloniki, Greece*, pages 1013–1016, October 2001.

[141] T. Hayashi, S.Yamasaki, N. Morita, H. Aida, M. Takeichi and N. Doi. Effects of IP Packet Loss and Picture Frame Reduction on MPEG1 Subjective Quality. In *3rd Workshop on Multimedia Signal Processing, Copenhagen, Denmark*, pages 515–520. IEEE Computer Society Press, September 1999.

[142] S. Gringeri, R. Egorov, K. Shuaib, A. Lewis and B. Basch. Robust Compression and Transmission of MPEG-4 Video. In *Proceedings of the ACM Multimedia Conference 1999, Orlando, FL, USA*, pages 113–120, October 1999.

[143] M. Chen. Design of a Virtual Auditorium. In *Proceedings of the ACM Multimedia Conference 2001, Ottawa, Canada*, pages 19–28, September 2001.

[144] S. Lavington, N. Dewhurst and M. Ghanbari. The Performance of Layered Video over an IP Network. *Signal Processing: Image Communication, Elsevier Science*, **16**(8):785–794, 2001.

[145] Developers – What Intel Streaming Web Video Software Can Do For You, Intel. http://developer.intel.com/ial/swv/developer.htm.

[146] Technical White Paper: PacketVideo Multimedia Technology Overview, PacketVideo. http://www.packetvideo.com/pdf/pv_whitepaper.pdf.

[147] O. Künzel. *Auswirkungen von Qualitätsveränderungen in Layer-Encoded Video beim Betrachter* (in German). Studienarbeit. Fachbereich Elektrotechnik und Informationstechnik, Darmstadt University of Technology, May 2002.

[148] Subjective Impression of Variations in Layer Encoded Videos. http://www.kom.tu-darmstadt.de/video-assessment/.

[149] R. Neff and A. Zakhor. Matching Pursuit Video Coding–Part I: Dictionary Approximation. *IEEE Transactions on Circuits and Systems for Video Technology*, **12**(1):13–26, 2002.

[150] J. Hartung, A. Jacquin, J. Pawlyk and K. Shipley. A Real-time Scalable Software Video Codec for Collaborative Applications over Packet Networks. In *Proceedings of the ACM Multimedia Conference 1998, Bristol, UK*, pages 419–426, September 1998.

[151] L. Vicisano, L. Rizzo and J. Crowcroft. TCP-like Congestion Control for Layered Multicast Data Transfer. In *Proceedings of the 17th Annual Joint Conference of the IEEE Computer and Communications Societies (INFOCOM'98), San Francisco, CA, USA*, pages 996–1003. IEEE Computer Society Press, March 1998.

[152] M. Garey and D. Johnson. *Computers and Intractability*. W. H. Freeman, San Francisco, CA, USA, 1979. ISBN 0-7167-1044-7.

[153] S. Podlipnig and L. Böszörmenyi. Replacement Strategies for Quality Based Video Caching. In *International Conference on Multimedia and Expo, Lausanne, Switzerland*, pages 49–52, August 2002.

[154] F. S. Hillier and G. J. Lieberman. *Operations Research*. McGraw-Hill, 1995. ISBN 3-486-23987-2.

[155] Ilog Cplex: Mathematical Programming Optimizer. http://www.ilog.com/products/cplex/.

[156] K. K. W. Law, J. C. S. Lui and L. Golubchik. Efficient Support for Interactive Service in Multi-Resolution VOD Systems. *VLDB Journal*, **8**(2):133–153, 1999.

[157] J. Kangasharju, F. Hartanto, M. Reisslein, and K. W. Ross. Distributing Layered Encoded Video through Caches. In *Proceedings of the Tenth Annual Joint Conference of the IEEE Computer and Communications Societies 2001 (INFOCOM'02), Anchorage, USA*, pages 1791–1800, April 2001.

[158] The Network Simulator ns-2. http://www.isi.edu/nsnam/ns/.

[159] M. Zink, C. Griwodz and R. Steinmetz. KOM Player – a Platform for Experimental VoD Research. In *Proceedings of the 6th IEEE Symposium on Computers and Communications, Hammamet, Tunisia*, pages 370–375. IEEE Computer Society Press, July 2001. ISBN 0-7695-1177-5.

[160] Realsystem Proxy8 Overview. http://service.real.com/help/library/.

[161] N. J. P. Race, D. G. Waddington and D. Shepherd. An Experimental Dynamic RAM Video Cache. In *Proceedings of NOSSDAV 2000, Chapel Hill, NC, USA*, June 2000.

[162] KOMSSYS. http://komssys.sourceforge.net/.

[163] NISTNet. http://snad.ncsl.nist.gov/itg/nistnet/.

[164] J. Schmitt, M. Zink, S. Theiss and R. Steinmetz. Improving the StartUp Behaviour of TCP-friendly Media Transmissions. In *Proceedings of the INC 2002, Plymouth, UK*, pages 173–180. University of Plymouth, July 2002. ISBN 1-84102-105-9.

[165] M. Zink. P2P Streaming using Hierarchically Encoded Layered Video. Technical Report TR-KOM-2003-01, Darmstadt University of Technology, January 2003.

[166] R. Tunk. Untersuchung von TFRC (TCP-friendly) *Mechanismen in drahtlosen Netzwerken der dritten Generation* (in German). Diplomar-beit. Fachbereich Informationstechnik, Elektrotechnik und Mechatron-ik, Fachhochschule Giessen-Friedberg, October 2002.

[167] R. Haskin and F. Schmuck. The Tiger Shark File System. In *IEEE 1996 Spring COMPCON, Santa Clara, CA, USA*, pages 226–231, February 1996.

[168] C. Martin, P. Narayan, B. Özden, R. Rastogi and A. Silberschatz. The Fellini Multimedia Storage Server. In Chung: *Multimedia Information Storage and Management*. Kluwer, 1996. ISBN 0-7923-9764-9.

[169] P. Corriveau, C. Gojmerac, B. Hughes and L. Stelmach. All Subjective Scales are not Created Equal: the Effects of Context on Different Scales. *Signal Processing, Elsevier Science*, **77**(1):1–9, 1999.

[170] QuickTime. http://www.apple.com/quicktime/.

[171] R. Koster, A. P. Black, J. Huang, J. Walpole and C. Pu. Infopipes for Composing Distributed Information Flows. In *International Workshop on Multimedia Middleware 2001, Ottawa, Canada*, pages 44–47, October 2001.

[172] H. Naguib and G. Coulouris. Towards Automatically Configurable Multimedia Applications. In *International Workshop on Multimedia Middleware 2001, Ottawa, Canada*, pages 28–31, October 2001.

[173] B. Li, D. Xu and K. Nahrstedt. Towards Integrated Runtime Solutions in QoS-aware Middleware. In *International Workshop on Multimedia Middleware 2001, Ottawa, Canada*, pages 11–14, October 2001.

[174] R. Vanegas, J. A. Zinky, J. P. Loyall, D. Karr, R. E. Schantz and D. E. Bakken. QuO's Runtime Support for Quality of Service in Distributed Objects. In *Proceedings of the IFIP International Conference on Distributed Systems Platforms and Open Distributed Processing (Middleware'98), The Lake District, England*. IFIP, September 1998.

[175] K. Mayer-Patel and L. Rowe. Design and Performance of the Berkeley Continuous Media Toolkit. In *Proceedings of SPIE/ACM Conference on Multimedia Computing and Networking (MMCN), San Jose, CA, USA*, pages 194–206, February 1997.

[176] T. Plagemann. *A Framework for Dynamic Protocol Configuration*. vdf Hochschulverlag AG an der ETH Zurich, Switzerland, 1996. ISBN 3-7281-2334-X.

[177] T. Kaeppner. *Entwicklung verteilter Multimedia-Applikationen* (in German). Vieweg Verlag, 1997. ISBN 3-528-05549-9.

[178] F. Eliassen and J. Nicol. Supporting Interoperation of Continuous Media Objects. *Theory and Practice of Object Systems: Special Issue on Distributed Object Management*, **2**(2):95–117, 1996.

[179] R. Becker. *Implementierung des Gleanings in die KOM VoD Umge-bung* (in German). Diplomarbeit. Fachbereich Elektrotechnik und In-formationstechnik, Darmstadt University of Technology, May 2001.

[180] E. Walthinsen. GStreamer – GNOME Goes Multimedia, April 2001.

[181] K.-A. Skevik, T. Plagemann, V. Goebel and P. Halvorsen. Evaluation of a Zero-Copy Protocol Implementation. In *Proceedings of the 27th Euromicro*

Conference – Multimedia and Telecommunications Track (MTT'2001), Warsaw, Poland, September 2001.

[182] S. Cen. *A Software Feedback Toolkit and its Applications in Adaptive Multimedia Systems*. PhD thesis, Oregon Graduate Institute of Science and Technology, Department of Computer Science and Technology, August 1997.

[183] *The IA-32 Intel Architecture Software Developer's Manual*, Volume 3: *System Programming Guide*. http://developer.intel.com/design/pentium4/manuals/245472.htm.

[184] F. Kon, M. Roman, P. Liu, J. Mao, T. Yamane, L. C. Magalhaes and R. H. Campbell. Monitoring, Security, and Dynamic Configuration with the dynamicTAO Reflective ORB. In *IFIP/ACM International Conference on Distributed Systems Platforms and Open Distributed Processing (Middleware'2000), New York, USA, 2000*, pages 121–143. IFIP/ACM, April 2000.

[185] J. S. Poulin. *Measuring Software Reuse: Principles, Practices, and Economic Models*. Addison-Wesley, 1997. ISBN 0-201-63413-9.

[186] R. D. Banker, R. J. Kauffman, C. Wright and D. Zweig. Automating Output Size and Reuse Metrics in a Repository-Based Computer-Aided Software Engineering (CASE) Environment. *IEEE Transactions on Software Engineering*, **20**(3):169–187, March 1994.

[187] W. Frakes and C. Terry. Software Reuse: Metrics and Models. *ACM Computing Survey*, **28**(2):416–436, June 1996.

[188] mpeglib.http://mpeglib.sourceforge.net/.

[189] MPlayer. http://www.mplayerhq.hu/.

[190] GStreamer. http://www.gstreamer.net/.

[191] R. Becker. *Design und Implementierung von Patching in die KOM VoD Umgebung* (in German). Studienarbeit. Fachbereich Elektrotechnik und Informationstechnik, Darmstadt University of Technology, September 2000.

[192] S. Theiss. *Konzeption und Implementierung eines Multiformat-fähigen Proxy Caches fuer die KOM VoD Umgebung* (in German). Studienarbeit. Fachbereich Elektrotechnik und Informationstechnik, Darmstadt University of Technology, June 2001.

Index

Scalable Video on Demand: Adaptive Internet-based Distribution M. Zink